D0501757

STUDIES IN BIOETHICS

WONDERWOMAN AND SUPERMAN

John Harris is Professor of Applied Philosophy at the Centre for Social Ethics and Policy in the University of Manchester. The Centre carries out research on issues of applied ethics and social policy. He is the author of *Violence and Responsibility* (1980) and *The Value of Life* (1985), and the editor of *Consent and the Incompetent Patient* (with Steven Hirsch, 1988) and *Experiments on Embryos* (with Anthony Dyson, 1989). His research includes work on the moral and political status of children, the just allocation of public resources, and the ethics of health care.

STUDIES IN BIOETHICS

General Editor *Peter Singer*

Wonderwoman and Superman
John Harris

The Values of Psychotherapy
Jeremy Holmes and Richard Lindley

The End of Life: Euthanasia and Morality
James Rachels

The Unheeded Cry: Animal Consciousness, Animal Pain and Science
Bernard E. Rollin

The Reproduction Revolution: New Ways of Making Babies
Peter Singer and Deane Wells

Wonderwoman and Superman

The Ethics of Human Biotechnology

John Harris

Oxford New York
OXFORD UNIVERSITY PRESS
1992

Oxford University Press, Walton Street, Oxford OX2 6DP
Oxford New York Toronto
Delhi Bombay Calcutta Madras Karachi
Petaling Jaya Singapore Hong Kong Tokyo
Nairobi Dar es Salaam Cape Town
Melbourne Auckland
and associated companies in
Berlin Ibadan

Oxford is a trade mark of Oxford University Press

British Library Cataloguing in Publication Data
Data available

Library of Congress Cataloging in Publication Data

Harris, John, 1945–
Wonderwoman and Superman: the ethics of human biotechnology/
John Harris.
p. cm. — (Studies in bioethics)
Includes bibliographical references.
1. Human reproductive technology—Moral and ethical aspects.
2. Genetic engineering—Moral and ethical aspects. I. Title.
II. Series.
RG133.5.H38 1992 176—dc20 91–23939

ISBN 0–19–217754–0

Typeset by Best-set Typesetters Ltd., Hong Kong
Printed in Great Britain by Bookcraft (Bath) Ltd.
Midsomer Norton, Avon

Acknowledgements

It will be obvious to anyone reading this book that I am not a scientist. I have not tried to become one, that would truly have been a lost cause. However, the issues discussed in this book are of such immense importance and interest that every citizen needs to become acquainted with them and to form views about what should be done or permitted to be done in the field of human biotechnology. Accordingly I have tried to acquire a level of scientific knowledge appropriate to such an understanding.

I have had one or two advantages in the attempt to do so. In particular I have benefited from the expert and patient tutoring of some of the best scientists in this field. These include Dian Donnai, Robert Edwards, Mark Ferguson, and Susan Kimber, I am very grateful to all of them for their quite invaluable advice. I am, however, a slow and sometimes reluctant learner and so despite such help I have doubtless failed to understand many things and consequently will, I am sure, have made some errors. I have of course made strenuous efforts to check or have checked all the scientific and technical facts in this book, but as I say, doubtless some errors will remain.

The knowledge required for an understanding of the problems created by biotechnology is not simply scientific, there is an important legal dimension to all this and Margot Brazier has been an unfailing and stimulating source of information and ideas.

Some friends have read and commented upon specific chapters or parts of chapters. I am particularly grateful to Margot Brazier, Susan Kimber, Søren Holm, Harry Lesser, Graham Macdonald, and Hillel Steiner. Others from whom I have learnt in discussion of these issues are Peter Braude, Len Doyal, Tony Dyson, Charles Erin, Max Elstein, Raanan Gillon, Sarah Warwick, Simon Winner, Bob Williamson, and Edward Yoxen.

Versions of Chapters 2, and 4 were originally published as: 'Embryos and Hedgehogs: On the Moral Status of the Embryo', in Anthony Dyson and John Harris (eds.), *Experiments on Embryos* (Routledge & Kegan Paul, London, 1990); and as 'The Wrong of

Wrongful Life', in Len Doyal and Leslie Doyal (eds.), 'Legal and Moral Dilemmas in Modern Medicine', *Law and Society*, 17: 1 (1990). I have, however, revised and rewritten both these pieces for this book.

Peter Singer and Susan Kimber have read the completed manuscript carefully and I am much indebted to them for useful comments and suggestions throughout, as I am to Edwin F. Pritchard for exemplary copy-editing.

Finally I want to record again the debt that I owe to Ronny Dworkin and Jonathan Glover, who continue to make me believe that philosophy can make a real contribution to the understanding and resolution of moral dilemmas.

Contents

Introduction

The rapidity of technological advance and in particular biotechnological advance has left our moral categories in disarray.

There was a rather poor joke which ran: 'motherhood is a fact, paternity merely a hypothesis'. Thanks to biotechnology motherhood is now also merely a hypothesis. The very concept of motherhood has been split in two by advances in *in vitro* embryology. These have made possible a distinction between the genetic mother, the mother whose gamete has been used and whose DNA has contributed 23 chromosomes to the resulting child, and the gestational mother who carries and gives birth to the child. There is of course also the possibility of a third and distinct social mother who does the mothering and brings up the child.

Similarly, birth has operated as a traditional divide. In law it marks the distinction between the offences of abortion and infanticide and most people celebrate birth as the arrival of the new human individual. We may be on the verge of an era in which ectogenesis, the nurturing to term of the fetus in an artificial womb, will be possible. When this happens the notion of birth will lose much of its legal and social significance.

At the moment we operate with the idea that it is wrong to damage the fetus in the womb, but should we take steps to improve it? Is there a morally relevant distinction between improving the embryo by genetic engineering and doing so by, say, modifying the diet of its mother during pregnancy or by educating the child later in its development?

In a sense these are the least interesting of the possibilities opened up by biotechnology. We are now able to transcend the limitations of particular species and combine the virtues (and vices) of different species and indeed programme into species new attributes never before a feature of any species. We can, or eventually will be able to, create new 'transgenic' creatures of unprecedented nature and qualities.

It would not be an exaggeration to say that humanity now stands at a crossroads. For the first time we can literally start to shape not only our own destiny in terms of what sort of world we wish to create and inhabit, but in terms of what we ourselves wish to be like. We can now, literally, change the nature of human beings. Whether we should do so and in what ways is the subject of this book. Should we celebrate the ingenuity and imagination of the biotechnologists who have made all this possible or should we rather try to limit and control their activities? This book will try to determine what our policy should be towards the possibilities opened up by biotechnology.

I shall not try now to review these possibilities in detail—this task will be undertaken in Chapter 1. I should now, however, make clear that this is a book about the ethics of human biotechnology. There are a wide range of issues in biotechnology that are principally about genetically engineering and modifying other life forms from animals and insects to plants and cells. In this book I shall be interested in these developments only in so far as they are used directly in the process of human genetic engineering. In particular I shall not discuss things like the release into the environment of genetically modified organisms and micro-organisms, the biotechnology industry and its social and economic impact, the use of things like animal growth hormone in farming and animal welfare issues more generally, pollution control and the treatment of waste products, biological warfare, the use by the first world of biotechnologically modified raw materials 'stolen' from the third world, and the international regulation and control of all these things.

This is an arresting list of topics that will not be covered. It is a great advertisement for someone else's book or perhaps for another and different book of mine. However, if you read as far as the next chapter you will see that there is plenty left to fascinate and to trouble us. The ethical and social problems raised by what it is already possible to do by way of modifications to human individuals and to the human species are important enough to warrant detailed and careful examination on their own. I hope this is what happens in the chapters that follow.

One aspect of specifically 'human' biotechnology has been deliberately neglected. This is high technology as applied for example to patient care in medicine generally. This is because although high-tech patient care is a reasonable extension of what one might understand by 'biotechnology', there is already extensive discussion of the ethical problems arising from the medical use of technology *per se*, in the now voluminous literature in health care ethics. I believe the essence of biotechnology is not 'the use of technology in the biosphere' but 'the use of technology in the application of biological science'.

As I shall be using it in this book, the term 'biotechnology' applies broadly to *the use of technology in the application of the biological sciences, especially genetics*.[1] This sense is, I believe, central and it is not inconsistent with three authoritative contemporary definitions of biotechnology. This may not of course accord with everyone's understanding of the term but it is, I think, both clear and paradigmatic. With the qualifications just noted it captures both the concerns of this book and, I believe, what is generally understood by the term 'biotechnology'.[2]

In considering the ethics of human biotechnology, I have always tried not simply to review the ethical problems raised but, where this is possible, to argue for definite conclusions about what should be done and what should be permitted to be done. Of course, I do not think that the conclusions I have reached are unassailable, nor that others will necessarily share some of my optimism about what could be achieved. I believe however, that I have produced arguments with which others can engage.

Despite setting out clear ethical conclusions over a wide range of the problems of human biotechnology I have no general view as to how access to genetic engineering should be controlled, nor as to how those who do the engineering should be controlled. There are many reasons for this. In the first place the ways and means of regulation in this field are immensely complex and would require and deserve a book-length study on their own. There is for example a complex debate about the rival merits of regulation by external bodies of various sorts on the one hand or the desirability of self-regulation by the various professions involved on the other. I imagine also that because of the sensitivity of issues in human biotechnology, each jurisdiction will want to formulate regulatory legislation in its own way; and that this legislation will arise from very different consultative processes and traditions in each case. For instance, in the more intellectual civil law tradition exemplified by France there is now a national ethics committee to address these issues. The English speaking world is, on the other hand, dominated by the more *ad hoc* common law tradition of 'judge-made law'.

What I have tried to do is discuss the sorts of moral considerations which should weigh with any individual or committee or legislature considering how, or indeed whether, to control human biotechnology. And I have tried to show, where possible, what precisely should weigh with them and how heavily it should weigh.

I should also perhaps give warning that I have tried to look beyond the end of my nose or rather humanity's nose in more senses than one. I have tried to anticipate the sorts of thing that may become possible, not next year but perhaps within the next twenty years. Of course I may be wrong about what will turn out to have been possible, but this is not important. I am not trying to predict nor yet to foresee. I am trying simply to look at the sorts of things that may be possible and to examine the arguments for and against doing those sorts of things. I have taken what I believe to be some of the best available advice about what might be possible but again this is not the most important point. What matters is what we should try to achieve, not whether we will in the end be successful.

Many people think looking too far into the future is irresponsible or frivolous. Irresponsible because such a process is inevitably error prone; and frivolous because there is enough to worry about now without inventing new things to worry about.

I believe strongly that both these approaches are mistaken. If we are to have any hope at all of retaining or perhaps gaining control of our own destinies, we must try to anticipate what might happen and form a view about whether we should be pleased or sorry about any impediments in the way of its occurring. We need to know whether we should try to increase the number and the strength of those impediments or do our best to reduce them. Of course we should worry about real and present dangers first. But future and possible dangers have an unerring habit of becoming real and present ones. And when they do they may be more difficult to control.

Shakespeare's Brutus, wrestling with the problem of whether Julius Caesar was such a danger and consequently should be nipped in the bud argued thus:

> Fashion it thus; that what he is, augmented,
> Would run to these and these extremities;
> And therefore think him as a serpent's egg
> Which hatch'd, would, as his kind, grow mischievous,
> And kill him in the shell.[3]

Let's grant that serpents are dangerous, although of course most are not. What we need to know are which eggs will produce serpents and which will produce the antidote to snakebite and what we are, morally, entitled to do about either.

While this book is about the ethics of human biotechnology I have not here attempted to give any general introduction to moral philosophy, nor have I tried to outline in general terms my own basic approach to ethics.[4] I have tried rather to let the arguments speak for themselves. There is always a danger when labels are attached to philosophical positions for people to assume that if they reject a particular school of philosophy in general, or adhere to a different philosophical tradition or approach, they can safely ignore or reject arguments from another school of philosophy. But most philosophical schools are united by their demand for rational argument and for the justification of moral conclusions. What matters is the quality of the arguments, reasons, and justifications produced. Here I am interested simply in what can be said for and against using biotechnology in various ways. For although the tools available to biotechnology are of enormous power and efficacy, whether or not they will be used, and whether they will be used for good or ill, depends in large part on political decisions and these decisions will turn in large part on the power of the arguments for and against their being used in various ways.

It has been said that politics is the art of the possible. I rather think of it as the art of the permissible. We ought only to do what is possible *if we ought to do it*. And in order to know whether or not we ought to do it, we need to know the arguments, and the quality of the arguments, for and against.

We are on the brink of a new revolution of quite awesome power. The revolution in molecular biology will give us the ability to divert and control human evolution to an unprecedented extent. It will enable us to manufacture new life forms to order, life forms of every sort. The decision before us now is not whether or not to use this power but how and to what extent. It might be tempting to pretend the revolution had not happened and to try to go on as before, but to do so would not only be futile, it might also involve us in causing an immense amount of avoidable suffering. There is no safe path. If we fail to make changes to human beings, the result may simply be that we ensure that the future will be much worse for everyone than it need be. If we make the wrong changes the same may be true. What we must try to do is learn to choose responsibly, but there is no sense in which doing nothing is necessarily a more responsible choice than doing something.

1

The Art of the Possible

Since some of the most interesting and important developments in biotechnology fall within the general area of genetics, we should perhaps say just a preliminary word or two about the core of genetics, namely the gene, so this is where we shall start. This chapter will then review the possibilities opened up by human biotechnology more generally. Subsequent chapters will largely concern themselves with deciding which of these things we ought to do or attempt, and which there are good arguments to abjure.

I. Genes

D. J. Weatherall gives a brilliantly concise description of the basic science: 'A gene is the unit of heredity that determines the structure of a peptide chain, that is a string of amino acids which form the building blocks of all enzymes and proteins . . . Every cell contains the genetic information to make an entire human being' and all 'this information is carried in deoxyribonucleic acid (DNA), the "spiral staircase" molecule described by Watson & Crick about 30 years ago'.[1]

Weatherall, who is Nuffield Professor of Clinical Medicine in the University of Oxford, goes on to give a dramatic picture of the scale of the problem posed by the science of genetics:

> In each cell in the body there is about 2 meters of DNA, tightly packaged and coiled. Since there are approximately 3×10^{12} cells in the body, if all the DNA from one human being was joined end-to-end it would stretch to the moon and back about 8000 times. When we are embarking on a search for a single gene of, say one or two thousand bases we are looking for a needle in a molecular haystack which is about 6 million times the size of the needle. It is a tribute to the remarkable advances in recombinant DNA technology that such searches are now almost routinely successful.[2]

Mapping the genome

The human genome is simply the complete set of genes of human beings. It is estimated that there are 3,500 million base pairs comprising the human genome. It is already technically possible to begin the task of mapping the human genome and determining the sequence of the 3,500 million base pairs required to complete the task. All that is needed is the resources. The cost is currently estimated at a mere 50 pence per base pair and the task could be completed in from three to thirty years.[3]

Why bother?

Gene probes

Well, apart from the sheer thrill of discovery, there are many beneficial possibilities. For one thing a very large number of the most serious human disorders are the result of a genetic abnormality. The pre-natal diagnosis of such disorders so that parents can decide whether or not to go ahead and have children who may or will be affected has for long been a high priority. Initially a combination of prediction on the basis of detailed family histories coupled with chromosomal analysis and chemical tests were the mainstay of such diagnosis. More recently, genetic probes have become available for the detection of many genetic defects in the embryo, Huntington's chorea for example. The pre-natal diagnosis of conditions such as Down's syndrome and spina bifida is already well established and, where they are detected, parents can be offered termination of pregnancy. While not regarded as a single gene disorder it is likely that a number of genes are in fact involved in the closure of the neural tube and consequently genetic screening might well pick out individuals predisposed to spina bifida. More and more probes for specific genetic defects are becoming available and pre-natal diagnosis of genetic defects will soon be standard and reliable.

Single cell biopsy

More important perhaps is the concurrent development of diagnosis using the DNA from a single cell. This single cell biopsy could become the basis of all predictive screening and opens up two different strategies for prenatal diagnosis.

First, using *in vitro* fertilization techniques, people at risk of genetic disorders, either because of a family history, age, and so on, could have embryos screened at the pre-implantation stage so that only healthy embryos would be implanted. Secondly, when gene therapy techniques become available it will be possible to correct genetic defects in the embryo, either at the pre-implantation stage *in vitro* or *in vivo* during pregnancy. Gene therapy techniques are at present some way off so for the moment we merely note this as a future possibility.

Genetic susceptibilities

As a picture of the human genome emerges, work will also progress in mapping particular genes to particular regions of the genome; and along with the identification of genetic defects will be identification of genetically determined susceptibility or resistance to certain diseases and increased knowledge about how all of this is affected by the environment. Such susceptibilities might include the likelihood of developing adult cancers, resistance to particular infections, and so on. The prospect opens up of a huge improvement in preventive medicine. For each individual could have her genome mapped as an embryo or at birth. This would not only reveal genetic defects which could be corrected as techniques emerge, but would also reveal genetic susceptibility to certain diseases. Here again, two different sorts of strategy might be indicated.

One would involve health professionals devising preventive strategies tailored to individual needs. Once the genetic defects, if any, and genetic susceptibilities of the particular individual were known she (or initially her parents) could be counselled as to the lifestyle appropriate to minimizing the chance of succumbing to conditions to which she had been shown to be genetically susceptible.

Alternatively, *in vitro* fertilisation techniques could be used for such screening with a view to implanting only optimal embryos. As Mark Ferguson, a prominent cell biologist, has remarked:[4]

> Different embryos will have widely differing profiles of susceptibility or resistance to various disorders when such widespread screening is applied. Therefore decisions about which embryos to reimplant will be all the more complex.

Ferguson continues by commenting upon the possibility of using such techniques to select for particular types of individual so that, for example, a particular society, government, or 'mad dictator' might try deliberately to create an élite group.

> Indeed it seems likely that the number of embryos fertilised in vitro would be the rate limiting step in any kind of purposeful selection experiment. The variation between embryos is likely to be so great that one would have to screen thousands, if not millions, of embryos from individual couples if one were to attempt deliberately to select certain consistent genotypes on a population basis. I therefore have no fears about the potential to manipulate the human gene pool (for good or bad); the sheer numbers required to achieve this precludes it as a rational possibility. I also believe that natural selection has left much to be desired, in terms of the human gene pool, and individual selections at this level are unlikely to have any detrimental effects, only advantageous ones.[5]

Ideal adults

While manipulation of the human gene pool to create a super race may be fanciful, fears about ideal individuals of various sorts may be more realistic. The ethical question of whether or not these fears ought to be fears or hopes I will for the moment postpone. However, cell biopsy and genetic probes might be used not only to screen for disease and disease susceptibility, but also for adult phenotypes such as gender, height, build, and perhaps also such traits as intelligence, longevity, and attractiveness.

Now of course screening for gender is nothing new. Selective infanticide has been practised in various societies for millennia and is still practised in our own times and sometimes in our own society. Moreover, amniocentesis may reveal gender and a mother might decide to terminate on the basis of such information. However, gender determination is likely to become relatively easy in the future by selecting particular sperm for fertilization. Techniques are already developed for use in animals which involve identifying and separating male and female determining sperm. The process is then relatively simple using artificial insemination *in vivo*.

The spectre raised by selection for gender is usually that of parents playing God and determining the sex of their children and perhaps even the ideal order in which to have children: boy–girl–boy and so on. Wider concerns concentrate on fears that some societies would favour the production of male children so highly that discrimination against females would be of epidemic proportions. Such policies would be likely to be self-correcting in the long term but an interim of profound misery might ensue. We should, however, remember that because a number of genetic disorders are sex-linked there may be compelling reasons to choose a particular gender for one's child in the absence of other therapies.

Screening for phenotypes other than gender is still some way off but the problems that such possibilities raise cannot be postponed. We will of course be returning to all these issues in due course.

Ideal children

Ideal children are perhaps those who are most likely to grow into ideal adults. However, the first ideal thing a child can do is come when, and in the state in which, she is wanted. In the euphoria following the birth of the first test-tube baby, Louise Brown, in 1977, everything seemed possible. One possibility widely canvassed was that of creating banks of good embryos.[6] Using cryopreservation techniques (freezing) a woman could produce a number of eggs while in peak reproductive condition, have them fertilized by her partner (or indeed anyone she fancied), and have the resulting embryos screened for any possible genetic disorders and then frozen and banked. She could then pursue her career knowing that, as she aged and became more susceptible to producing eggs with chromosomal anomalies or genetic abnormalities, she had a healthy credit balance in the bank which she could withdraw when desired and implant with confidence.

This rather clinical and conscious control of nature for one's personal convenience and priorities has produced much hostility. Again the respectability of such hostility is something to which we will return. Recently, some doubt has been cast on the scientific viability of this scenario. Again, Mark Ferguson:

> However, it is well known that the maternal environment, e.g. the amount of blood flow to the uterus, the nutrition of the embryo, the volume of amniotic fluid etc. all play a major role in successful fetal outcome. It is unclear how well the ageing human female reproductive system would cope with such good embryos. For certain they would not have the major malformations like Down's syndrome, but perhaps they would be less than optimal in their development. I would therefore only see such techniques being applied when there was some pressing need for the mother to be pregnant in later years . . .

In no way would one advocate such procedures for the routine delaying of pregnancy e.g. until the mother had achieved a suitable prominence in her career. Such widespread voluntary practices would be contraindicated a) because of the likely poor environment for the embryo in the ageing female reproductive system and b) the simple fact that the older you are when you produce children the less time you are likely to have (in terms of life expectancy and quality of life) to bring them up to an age at which they become independent.[7]

Whether or not pregnancy is delayed, the embryo bank offers other advantages. As R. G. Edwards has noted:

Identifying embryos with genetic abnormalities would offer an alternative to amniocentesis during the second trimester of pregnancy, and the 'abortion *in vitro*' of a defective preimplantation embryo, still free-living, minute and undifferentiated, would be infinitely preferable to abortion in vivo at twenty weeks of pregnancy or thereabouts as the results of amniocentesis are obtained. It would also be less traumatic for parents and doctor to type several embryos and replace or store those that are normal rather than having the threat of a mid-term abortion looming over each successive pregnancy.[8]

Before leaving the general topic of ideal children we must just mention the possibility that cryopreservation opens up of *post mortem* conception and birth. Either frozen embryos or separately frozen eggs and sperm may be thawed and the embryo implanted after the death of one or both of the 'parents'. In a recent case a French woman had to fight through the courts for the right to obtain the frozen semen of her dead husband so that she could have their child after his death from cancer.[9]

It is time to introduce a basic distinction to which we will refer repeatedly in the course of our deliberations. It is that between manipulations of the 'somatic line' and those of the 'germ line'. Briefly, operating on the 'somatic line' is limited to the cells of a particular individual and would not become part of its transmittable genetic make-up and so would not be inherited by children of that individual. Operations on the 'germ line' on the other hand alter the genome of the individual and its offspring. These latter manipulations are particularly important because they open up the possibility of creating permanently modified and even new and unprecedented life forms.

Operating on the somatic line

Manipulating the somatic line can be important for a number of reasons. Genetic disorders detectable in the embryo might be corrected by somatic gene therapy of various sorts. For example if a particular gene were missing in an individual this might be replaced. Techniques already exist for inserting particular pieces of DNA in to the correct stem cell line and this technique may well be effective in preventing major handicaps. Equally, somatic line interventions may take place at a cell level rather than involving modifications to the particular embryonic genome. Where anomalous cells were detected in the early embryo these might be removed 'either surgically or by killing them using toxins coupled to specific monoclonal antibodies which specifically recognised those abnormal cells'.[10]

Ferguson believes the introduction of additional cellular material to be the more likely technique.

> Thus, for example, if the embryo were deficient in cells making a particular type of hormone or particular blood cells, then stem cells for such components could be introduced into the embryo to correct the defect ... It seems likely that most somatic line cellular manipulations would occur in later embryos or fetuses.[11]

Operating on the germ line

The crucial advantage of manipulating the germ line of an individual is that any modifications thus made become part of the genome of that individual and will be inherited by its offspring. Moreover, significant modifications to the germ line of an individual alter its permanent genetic make-up and hence, in a sense, its identity. Creatures so altered have been held by United States courts to be new life forms and have even been patented as such. The first such creature to be patented was the now famous 'oncomouse', a mouse genetically altered to readily express the *myc* oncogene and hence be readily susceptible to the development of tumours. The oncomouse has been widely heralded as a valuable tool in cancer research.[12] These new life forms can reproduce and so establish themselves in the world. We should note for the record that while some genes inserted in this way become permanent parts of the genome capable of indefinite transmission to endless generations of offspring, other genes (so called 'unstable inserts') are lost after a certain number of generations. One consequence of all this, however unlikely its actualization may seem at the moment, is that the possibility is opened up of modified human beings becoming a distinct new species, perhaps in competition with more conventional humans. Here of course I have dashed headlong into a science fictional future, a tendency which it is all too difficult to curb when contemplating biotechnological advance. However, in this field we are already in a sense in the middle of a science fictional future so let's just see what it looks like.

Transgenic animals

One method of altering the genome of an individual animal or plant is to include parts of the DNA of another organism. This can of course be done by traditional breeding methods and such deliberate modification of species has probably been done for millennia. However, here we are talking about achieving transgenic animals or plants by methods other than breeding. Usually, non-parental DNA is introduced by injection into one of the pronuclei of the early embryo. This can easily be done with mice for example but the process can be more difficult with larger vertebrates, although it has proved feasible with domestic animals like pigs and sheep.

Other methods being rapidly developed at the moment involve either using modified retroviruses which can integrate a DNA copy of their genome into the chromosome of their host and hence deliver non-parental DNA into the embryo's chromosomes, or using very early embryonic cells which can be introduced into another early embryo. These early cells are 'toti potential', which means that they have the potential to become any part of the developing organism. That is to say they are not at this stage 'earmarked' to become, say, ears or, for that matter, brain cells, but possess the potential to develop into any part of the creature which will evolve from them.

Now these methods promise much. For one thing they may well cast light on the process of development of the individual from single cell to complete organism. This is not only interesting in itself but may well explain how this process sometimes goes wrong and so may provide means of correcting aberrant development of the organism. In particular, for example, it may explain why and how certain types of cells only develop in particular regions of the body or demonstrate which cells in the adult are derived from which regions of the embryo, or how regions of the body composed of identical tissue come to have different patterns of organization, like the hand and the foot, or how those genes develop which are involved in cancer (oncogenes).

More prosaically but equally important, new organisms can be vital in both the medical industry and indeed in agriculture. I cannot improve upon Mark Ferguson's elegant summary of these developments, which

> include the ability to produce pharmacologically important compounds in novel, easily accessible body compartments e.g. the production of clotting factor nine in the sheep mammary gland where it can be extracted from the milk, the production of organisms genetically resistant to disease (to overcome problems of vaccination or more importantly diseases where no vaccine exists), to develop more efficient domesticated livestock e.g. to get animals to grow bigger, faster and at higher food conversion ratio without upsetting the physiological balance: (such procedures have been achieved by engineering growth hormone releasing factor genes to be overexpressed in various animals), to produce animals capable of existing in an adverse environment (e.g. by engineering in genes which confer resistance to certain environmental pollutants such as heavy metals) and to develop better animal models for human disease (e.g. a better mouse model for mammary gland cancer) or to develop animal models for human diseases where no comparable disease exists in animal populations (e.g. muscular dystrophy or Lesch-Nyan syndrome).[13]

Transgenic humans

The techniques which have already produced transgenic animals may as well be used for a number of human purposes. If human embryos for example had defective genes these might be replaced with appropriate healthy genes and the defect would be repaired not only for that embryo but would be a permanent repair inheritable by its own children in due course. This procedure is already feasible for the replacement of single genes but is likely soon to be feasible for multiple gene replacement where for example genes have not been expressed in the embryo or have been expressed in the wrong place or in incorrect numbers.[14]

More fascinating by far is the possibility of inserting additional genes into human beings. This procedure would not be to correct genetic defects but to insert brand new genes—genes which do not naturally or normally occur in humans, creating a genuinely transgenic being. The point and the justification of this procedure would not be to repair genetic defects or cure dysfunction but rather would be to enhance function. It should be emphasized that this procedure would be highly dangerous and problematic at present because of the possibility of the new genes or gene products destroying the delicate balance of the existing genetic make-up of the individual with unforeseen consequences.

This distinction between repair and improvement, between removing dysfunction and enhancing function is deeply felt and is one which will occupy us at some length as the discussion proceeds.

Wonderwoman and superman

Transgenic humans are most likely to be produced by operating on individual embryos produced by *in vitro* fertilization, although it may be possible genetically to modify the gametes to get the same results or perhaps even do genetic surgery *in utero*. All these possibilities will be discussed later.[15] Although such modified humans would be able to pass on their modification by sexual reproduction thereafter, they would have to breed exclusively with other transgenic adults in order to be sure of maintaining the required characteristics. These constraints would certainly limit the scale on which transgenic humans could be produced and the scope for abuse. Again, the dangers and the ethical dilemmas posed by these possibilities will be addressed later.[16]

For the moment we will remain content with simple scene setting and just rehearse the senses in which these transgenic men and women might be considered 'super' and 'wonder'.

The benefits of producing some transgenic humans with enhanced function are formidable, and these benefits might accrue to the society as a whole as well as to the individuals themselves. These benefits are again admirably and concisely summed up by Mark Ferguson:

> It may be possible to insert individual genes coding for antibodies against all the major infections including hepatitis B, malaria, AIDS etc, or to insert genes which code for enzymes which would destroy carcinogens or environmental pollutants, or for genes which would repair DNA and so retard ageing, or even engineer in biosensor genes. In the latter technique one could insert a gene which would cause a colourful protein to be excreted in

the urine once the level of a particular molecule within the body (e.g. an oncogene protein which is precancerous) reached a certain threshold level. In other words, when your urine turned green, you would know that you were about to develop a malignant tumour in your lung and could go straight away to the hospital for early corrective treatment of that cancer.

Such positive gene therapy may be extremely useful for protection of certain individuals against various environmental pollutants and major new infections such as AIDS.[17]

A key to heart disease?

Perhaps the greatest of the contemporary killer diseases is heart disease. Here too there may be hope of preventive manipulation at the cellular level. As D. J. Weatherall has commented, 'Undoubtedly therefore, there is a strong genetic component to the development of premature vascular disease'.[18] Weatherall goes on to speculate about various ways in which recombinant DNA technology may provide methods of, for example, controlling the damaging effects of cholesterol. If the promise is fulfilled then one of the geatest killer diseases of the contemporary world may be put into retreat. Again it must be emphasized that these developments are still some way off and of course may never be realized.

Wound healing

Still on the cell level but contemplating the possibilities not so much of manipulation at the embryonic stage, but on developing more general treatments, it may be possible greatly to accelerate wound healing and reduce subsequent scarring by the external application of engineered molecules or clones of cells. The future may well hold the routine application of, say, biological surgical dressings rather than the more conventional 'band aid'. Moreover, if scarring can be dramatically reduced this may have important psychological advantages and may reduce the necessity for subsequent plastic surgery in severe cases.

With promising lines of research opening up for the treatment of cancer, heart disease, AIDS, and other major infections, fundamental changes may be predicted for medical education and research. Weatherall believes that this will result in a shift from preoccupation with the 'whole patient and individual organ pathology to the study of disease at the cellular and molecular level ... This will require a complete change of thinking about how our academic departments are organized. And it will require careful rethinking about how best to integrate the basic and clinical sciences.'[19]

Woman's world

A curious, but by no means insignificant, by-product of the massive work that is progressing in human embryology is the possibility of artificial parthenogenesis. This is the process whereby the unfertilized human egg (and indeed the eggs of other animals and mammals) can be stimulated to grow without fertilization. The result is a female individual who is a near clone of the mother. It is believed that parthenogenesis occurs randomly but rarely in most species and with eggs harvested for *in vitro* fertilization the process can be induced by the simple expedient of dipping the unfertilized eggs in alchohol.[20] This possibility will be discussed in more detail in the next chapter but we should note at once that there is some evidence that parthenogenetically stimulated embryos are unlikely to implant successfully. Whether this problem can be overcome making parthenogenesis a practical and not merely a theoretical possibility is as yet unknown. I have discussed the moral, political, and social consequences of this procedure at length elsewhere[21] but for the sake of completeness here we should note the, as yet theoretical, possibility that this opens up of a viable all female society.

If a group of women with the appropriate technological resources so chose, they could form a viable exclusively female society, with the capacity to reproduce without any recourse to men at all, even in the humble role of sperm donors or 'breeding bulls'. Moreover, individual women who wanted children but who were resolutely opposed to associating with men could, using artificially stimulated parthenogenesis, have children of their own, their *very own*. Indeed this latter consideration might prove highly attractive, since most parents want their children to be like themselves, the prospect of producing a near clone might be irresistible to some, particularly when their child's genetic make-up would be entirely their own. Of course they could only have females by this process, but, presumably, such women would not regard this as an important disadvantage of the procedure.

Genetic fingerprinting

In 1985, Alec Jeffreys and his team in Leicester developed a gene probe which reveals regions of the DNA which display patterns unique to each individual.[22] This probe, now commonly called the 'Jeffreys probe', has a variety of uses. For one thing, given a minute biological specimen, blood, saliva, semen, etc., the Jeffreys probe can be used to determine from which individual it comes by matching it to that individual's DNA as revealed from a similar specimen—hence the shorthand 'genetic fingerprinting'. This probe has a number of uses. It can accurately determine paternity or maternity, it can be used to establish whether or not an individual is genetically related to others and has been used in this application to determine family relationships for immigration purposes. It can be used to determine the identity of rapists and other criminals who have left even the most minute biological specimens behind. It also plays a crucial role in determining susceptibility to disease, which we have already considered. Some doubts have been cast on the effectiveness of this technique in criminal cases in the United States but the Jeffreys probe is still widely regarded as reliable.[23]

These applications are of course highly commercial and ICI have recently acquired the patent to the Jeffreys probe and have set up a company called 'Cellmark Diagnostics' to exploit the commercial possibilities of genetic fingerprinting.

Intellectual property

One interesting issue of general and very wide significance which is highlighted by the biotechnological explosion, but by no means peculiar to it, is that of the ownership of ideas and techniques, so called 'intellectual property'. This problem has many dimensions which we can simply signpost at this stage.

The first is the philosophical question concerning the ownership of such things as ideas. Are they the sorts of entities that can be owned and what theory of ownership is appropriate? Second comes the question that is not simply legal but also both philosophical and theological: can and should new life forms be patented? Can we not only play God, but purport to usurp her legal title? Finally there is the wider ethical issue of whether or not ideas and techniques which are of such major public importance and of such potential benefit and danger can, or ought to be, privately owned at all; or whether they must begin and remain in the public domain?[24]

II. Embryos

Way back in 1981 Robert Edwards reported[25] that 'it is now possible to contemplate the use of "tailor made" embryonic tissue grown in vitro for grafting into adults'. Noting the advantages mentioned above, Edwards continues, 'Grafts of embryonic tissue may offer a wider scope than those taken from neonates or adults, because tissue could be obtained from organs which do not regenerate in adults, and the risks of graft rejection can possibly be eliminated.'[26] Edwards goes on to list a number of the specific possibilities as they appeared in 1981. He noted that fetal tissue might be used to replace bone marrow in patients who had been exposed to radiation and fetal liver cells injected into the placenta might prevent the expression of inherited anaemia. A 'practical approach to controlling immunological ageing may involve a combination of dietary manipulation, chemical therapy and cell grafting ... Other recent reports have indicated that pancreatic cells may be used to repair diabetes and cultured skin cells grafted to repair lesions ... Human amniotic epithelial cells ... could be useful in repairing inherited enzyme defects in recipient children and adults.'[27]

Edwards further reported that there were indications that fetal brain tissue might be capable of repairing neural defects in adults and 'there are reports that kidney cells may be transplanted into the human brain in order to cure illnesses such as Parkinson's disease ... Myocardial tissue ... should be obtainable from embryos growing in vitro without great difficulty'[28] and might be used by cardiologists for repair of the major vessels of the heart.

Another prognosis rehearsed in 1981 was that it might be possible to avoid the rejection problems that were then bedevilling transplant surgery by 'tailoring embryos to suit a particular recipient'[29] and he listed two major advantages in using fetal tissue. The first is that such tissue 'might not be rejected by incompatible donors as strongly as adult tissue' and that 'Tissues compatible with an adult host might also be obtained through cloning' or otherwise by genetically tailoring matched or compatible tissue.[30]

The future is still with us

Most of what seemed possible in 1981 has now come about or is still clearly on the agenda for using embryonic material for therapeutic purposes. A decade is a very long time in biotechnology and this should be borne in mind when considering scenarios that seem merely science fictional today, including perhaps many that have been and will be rehearsed in this book. The first fetal transplants to treat Parkinson's disease have now taken place and it will not be long before the first test-tube babies are having babies of their own.

It is clear that embryo cells have certain special and highly advantageous features. Adult brain cells do not regenerate whereas embryonic cells do. This signals the possibility of using embryo cells to restore brain function and repair neural defects in adults. Indeed, it is now well established that embryo cells generally have an amazing capacity to invade and colonize adult organs, taking over or supplementing the function of these and so repairing damage to adults. Moreover, embryo cells and indeed embryonic tissue and organs are much less susceptible to rejection when transplanted than are adult to adult transplants. There are then special and quite major advantages to contemplating the use of embryo cells, tissue, and organs for transplant and grafting into adults. Indeed, animal experiments have demonstrated that embryo cells may be transplanted with success across major species barriers and recent work[31] suggests that it may be possible to transplant organs successfully across species barriers also. This possibility we will consider separately in a moment.

We can now look in more detail at the ways in which the embryo or fetus might be exploited. I use the term exploitation literally: the ethics of exploitation is an issue which will, of course, recur in later chapters. We can consider the ways in which the embryo might be used for beneficial or therapeutic purposes under four general headings.

1. Cloning

Clones have for some reason gripped the imagination. The idea of 'Xerox-copy' individuals of unlimited reproducibility is certainly arresting if not wholeheartedly attractive. Therapeutically, the idea of cloning is important because, as Robert Edwards's prediction makes clear, cloned individuals would share the same immune characteristics as each other. The prospect of, say, cloning an individual at the embryo stage so that one clone could be used as a cell tissue and organ bank for the other immediately arises. More remote but no less sinister perhaps is the idea of allowing one of the resulting 'twins' to develop normally—to be implanted and grow to adulthood, while allowing the other to develop as perhaps some sort of bank of compatible material, perhaps to be frozen or otherwise kept in suspended animation for use as and when the need arises.

We must start by making clear a distinction between the different procedures that might be described as 'cloning', in particular between cloning embryos and cloning cells or cell lines.

Single cells may be taken from embryos, fetuses, or even from adults and may be grown in culture. These may divide very many times, each cell being an identical clone of the original. These cells are useful for study but they cannot be grown into an embryo or adult creature. On the other hand, early embryos grown *in vitro* may be divided into two or even four separate embryos. Where this happens naturally twins or quads result. This process has an in-built limit and so the prospect of infinitely reproducing identical embryos is not possible by cloning embryos.

More dramatic is the prospect of deleting the nuclei of the cells of an early embryo and substituting the nuclei of cells taken from a different embryo or even an adult. This procedure, while technically very difficult to achieve, has given rise to speculation about the possibility of vainglorious adults reproducing identical clones of themselves by having nuclei removed from some of their own cells and inserted in those of *in vitro* growing embryos.

Clones produced by nuclear substitution will not be identical to the nucleus donor, but they will be very similar indeed. The reasons they will not be identical are many. First, a cell consists of two parts, the nucleus and the surrounding cytoplasm. The cytoplasm contains mitochondrial DNA and this may play a minor part in the genetic make-up[32] of the resulting individual; and since the cytoplasm is not substituted by this process its influence will partly shape the clone. Equally, the environment has profound influence on the developing individual. First, the environment *in utero*: we know how important it is that human mothers for example avoid smoking, alcohol, and ingestion of drugs during pregnancy because of the adverse effects of these practices on the embryo. Equally the embryo's nutrition depends on the mother's diet and will play a role in the development of the embryo; as will more abstract features such as the volume and composition of the amniotic fluid. Following birth, the physical and emotional environment of the child will also exert incalculable[33] influence on development. All this said, the clone will still be physically very like its 'parent' nuclear donor, and this might be enough to satisfy vanity, if that is the motive.

2. Cell banks

We have noted the special feature of embryo cells which enables them to invade and colonize adult organs and their markedly reduced susceptibility to rejection by the host immune system. This will make possible the creation of a pool or bank of cells derived from experimental or 'spare' embryos.[34] These cells would not need to be clones of the individual recipients, but would be compatible simply in virtue of the invasive and colonising capacities of embryo cells generally. Now these cells would be derived from embryos, probably so-called 'spare' or 'experimental' embryos produced as part of an *in vitro* fertilization programme. The distinction between 'spare' and 'experimental' embryos is of interest. Spare embryos are essentially normal embryos which are surplus to requirements for fertilization. They occur because candidates for IVF are induced to 'superovulate', that is, drugs are administered which stimulate the production of a number of eggs in a given cycle. These are 'harvested', fertilized, and the resulting embryos examined. Those that are healthy and normal will then either be implanted in the uterus of the mother, or may be frozen and stored for future use. If there are more than can be required for implantation then these are the spare embryos that may become candidates for use as banks of donor cells or for experiments. They appear highly eligible for experimental use because they are 'spare'—surplus to requirements.

Some researchers, notably Robert Edwards,[35] make a further distinction between these healthy and normal spare embryos and those which on examination prove defective or damaged in some way. These would not be implanted and are called by Edwards 'experimental' embryos to distinguish them from healthy spare embryos which could, theoretically, be implanted if there was a uterus waiting for them. The ethical respectability or otherwise of such cell banks will be considered in more detail later.

3. Non-viable neonates

It is possible to keep alive newborns with severe defects which would render them non-viable in the long term. Babies born with anencephaly for example—with no brain—would be alive at birth but could not long survive. They might, however, be kept alive long enough to become a useful source of organs, tissue cells, and so on. For many people, neonates like this come into a separate moral category and the case for using neonates as organ donors is a separate issue.[36]

4. Use of fetal material for transplants

This is likely to be one of the most important developments in the near future. Already the use of fetal tissue in the treatment of Parkinson's disease is making headlines and we have already reviewed many of the other expected uses. Here we need to note some of the different ways in which fetal and embryonic material might be used.

(*a*) *Embryo to embryo transplants*. In addition to the possibility of genetically modifying the fetus by the replacement or introduction of cells and genetic material to create transgenic humans which we have already considered, the embryo is a highly eligible transplant donor and recipient. Great success has been achieved in animals with the transplant of all the major organs, heart, kidneys, liver, etc. Such transplants are more successful in the embryo than in adults both because of the substantially reduced likelihood of rejection and because the embryo will grow to accommodate the transplants, unlike adults whose growth is complete. There is every reason to believe the same success can be achieved in human embryos, and indeed so-called fetal therapy is already one of the major growth areas in medicine. This therapy can often be performed on the embryo or fetus *in utero* with the added complication that this can only be done through the body of the mother. There is already case law in the United States in which a mother has been compelled to undergo a Caesarean against her will in the interests of the health of her unborn child.[37] The competing rights of mother and embryo will continue to be of great interest and controversy.

(*b*) *Embryo to adult transplants*. We have already reviewed the possibility of embryonic cell banks and indeed, of the transplantation of tissue and organs from specially engineered clones. There remains the case of the use of spare or experimental embryos as organ donors, and indeed that of aborted fetuses later in the cycle of development, as sources of material for transplant or grafts into adults.

In animals transplants of the major organs from fetus to adult have proved successful where the embryonic organ has been implanted in tandem with a failing adult organ, initially to augment function and to continue to grow and gradually to take over from a failing system. If the animal work can be repeated in humans, as there is every reason to suppose that it can, then there will be an increasing demand for fetus donor organs and fetal cells and tissue.

III. Animals, Plants, and Other Organisms

1. Animals

Animals figure in the development of biotechnology in two distinct ways. The first is as experimental subjects. In this role they may be used to test procedures which will eventually be used in humans, or to test the probable effects on humans of products that may be released into the atmosphere or used on animals or plants. They may also be used to test procedures which are destined to be used on a commercial scale for the modification of animal species.

The use of animals raises general questions about the ethics of treating animals in various ways and also about the issue of the exploitation by one species of, effectively, all the others. There are also more particular and sharply focused questions about the ethics of modifying particular animals or species or indeed creating and patenting brand new life forms.

There are also questions about the use of animals as a source of material for therapeutic use in humans. Animals have long been used in this role. The development of insulin for use in the treatment of diabetes has depended on the provision of animal and, in particular, pig insulin. A baboon heart has been used for transplant into a human recipient and work is progressing[38] in the United Kingdom, at Dulwich Hospital and elsewhere, on a technique for removing the antibodies which cause organ rejection in humans and so possibly paving the way for animal organs to be routinely used in transplants.

Experimental work on animals is already well under way. Human growth hormone has, for example, been engineered into mice to create a race of fast-breeder 'super mice' which also grow to at least twice the size of normal mice. Different species have been crossed, notably at Cambridge, where cells from sheep and goats have been fused to form a cross or 'chimera'; but whether or not a 'lake of fire' figured in the experiments is not recorded.

Already techniques of *in vitro* fertilization and embryo transfer are being used in animals. Cow embryos are routinely removed from the cow and frozen so that they may become a commercial product. The progeny of pedigree cows can in this way be sold world-wide, maximizing profits and indeed the number of progeny that can be produced by one pedigree mother. It is likely that recombinant DNA techniques will be used to breed super efficient animals with an enhanced food conversion ratio or ones that are more resistant to disease or able, as we have seen, to function and thrive in adverse environments.

Many people have misgivings about such control of nature. A leading American polemicist, Jeremy Rifkin displays an almost mystical unease recording his preference for traditional categories and for 'the notion of a species as a separate, recognizable entity with a unique nature or telos'.[39] These are interesting issues and ones to which we will return in due course.[40]

2. Plants and micro-organisms

The use of genetic engineering in plants and other organisms, bacteria for example, is now almost routine in commercial agriculture although open release of genetically modified organisms is still widely and strictly controlled. Plants and trees are cloned so that new or advantageous strains can be literally mass produced. In the long term it may be possible to engineer nitrogen-fixing plants. Such plants would effectively produce their own fertilizer and so would not need artificial fertilizer. They might also grow in very poor soil and thus literally 'make the desert bloom'.

The mass production of genetically engineered bacteria is also possible and may have numerous uses. I shall give just one example.

A very commonly occurring bacterium called P-syringae has the property of aiding the formation of ice crystals. It occurs on plants throughout the world. Scientists at the University of California have genetically modified this bacterium so that instead of aiding in the formation of ice crystals it has the reverse effect. Since a great enemy of crops is frost damage it might be possible to release this genetically engineered bacterium, called appropriately 'ice-minus', on a large scale, driving out naturally occurring P-syringae and replacing it with ice-minus, thus producing frost-resistant crops.[41]

Genetically engineered animals, plants, and bacteria pose special problems. They have three characteristics which make them uniquely different from other products of advanced technology. Like the germ line modifications considered earlier in this chapter, genetically engineered plants and animals pass on their modifications to future generations. Moreover, and more important, they can migrate. So it is impossible to confine them within controlled boundaries. Thus, once released or planted commercially, they cannot be recalled to the laboratory or eliminated from the environment.

The question thus arises: given the irrevocability of genetically engineered changes to bacteria, plant, and animal species, and given the rapid rate of reproduction and migration compared, say, with human beings, should we run the risks of permanent and irrevocable changes to the eco-system?

IV. Conclusion

There is much to talk about! And this we must now begin to do. The discussion will begin by setting some of the parameters to consideration of the ethics of human biotechnology and subsequent chapters will consider in detail many of the issues and developments we have just rehearsed.

This is a book about human biotechnology. We will therefore mention animals and plants only in passing, as they affect developments in human biotechnology. So far they have just crept into the discussion in their own right, but, I am afraid, this is as far as they will get. The issues that they raise are fascinating and I hope to discuss them in another book, but for now we must concentrate on the ethics of human biotechnology.

2

Research on Embryos

At the heart of human biotechnology has been the embryo. First, of course, because it was the study of the embryo and the ability to fertilise the human embryo in a glass dish on the laboratory table that led to the birth of the first 'test-tube' baby, Louise Brown, and to the explosion of general interest in human embryology. It is the development of the embryo, the way in which genes come to express themselves in particular ways, and the ways in which this expression might be modified or influenced, that still holds centre stage. In particular of course, the question of whether many of the possibilities we have just been examining might actually be realized depends in large part on the continued study of the embryo, and the human embryo in particular. This of course raises the question of the legitimacy of experiments on embryos.

Similar ethical issues are raised by the use (perhaps without 'manipulation') of embryonic cells, tissue, or other products, and indeed the use of neonates and aborted fetuses as sources of therapeutic or experimental material. These are the questions we must now examine.[1]

The question of whether or not we should permit research on or using the human embryo and of whether it is permissible to use the embryo or embryonic tissue for experimental or therapeutic purposes is about as vexed a question as one could hope to find. There are a number of perspectives from which one might approach such questions. There is for example a distinctive feminist approach which 'does not recognise the embryo as a separate human entity. It makes women and the social context central to its position'.[2]

Equally there is what might be called the strictly utilitarian approach which sees the issue simply in cost–benefit terms and wants to trade off the moral costs in terms of people's sensibilities about the treatment of embryos against the benefits to humanity that research will bring. Then there are those who see the issue simply as one concerning the rights of the embryos and regard the question as settled by the 'fact' that the embryo is a human being and so possesses 'human rights'.

Whatever the perspective from which one approaches these questions there is one unavoidable and central issue, and that is what we might call the moral status of the embryo. For what one might be entitled to do with or to an embryo will depend on its moral importance or status in precisely the same way as it does for the rest of us. The moral justification of the protections that surround you and me, in virtue of which we cannot simply be used or experimented upon without our consent, derive from the moral differences there are between us and other creatures and from the morally relevant features we possess. So that if the embryo is of a moral importance comparable with that of, say, normal adult human beings then the moral rights and protections possessed by such beings extend to the embryo. If it would be wrong to experiment on normal adults without their consent or to use them simply as tissue and organ banks, then it would be wrong to treat embryos likewise. So that when feminists say that they do not recognize the embryo as a separate human entity, this is a respectable position only if it is the conclusion of a moral argument which addresses the question of what status the human embryo should have.

So this is a first question that must be answered. But if we can answer it there may still be further considerations that properly bear on the permissibility of experiments on embryos and these we shall consider later. First then, can we determine the moral status of the embryo?

I. Embryos and Persons

Mary Warnock is a philosopher who has been enormously influential on the question of what it is permissible to do to and with embryos, both in her role as Chairman of the Warnock Committee[3] and in her many subsequent writings about embryos. In a recent major essay entitled 'Do Human Cells Have Rights?'[4] Mary Warnock addresses the central question before us now. She starts with the issue of whether or not the question of the moral status of the embryo is the same question as whether or not the embryo is a person. She notes that there is a natural presumption that 'if it can be shown that the embryo is a person, then it will follow that it has rights, for certainly all persons have rights, and, it is sometimes held, only persons have them'.[5] She thinks this approach mistaken for reasons it is as well to be clear about at the outset of the discussion.

Taking as her point of departure an earlier discussion of mine[6] on this point Warnock comments:

> John Harris of Manchester University . . . argued that to ask whether using human embryos should be permitted or not, and for how long, is to ask when human life begins to have moral significance. With this I would completely agree. But he goes on to say that this question is the same as the question when does an embryo become a person? And here I think that confusion is likely to set in. For the question about moral significance, the question, that is, when do embryos morally matter, is quite obviously one that must be answered by judgement and decision, according to a particular moral standpoint. It is not a question of fact but a question of value. How much should we value human life in its very early stages? But to translate this into a question about whether or not in its early stages an embryo is a person looks like translating this into a question of fact . . . That personhood, its possession or non possession, is as much a question of value as is the question when human life begins to matter, is hard for people to grasp.

With small cavils that are irrelevant to our present discussion I agree with what Warnock says here and indeed I was using 'person' in just this sense, as a shorthand term for all the reasons we have for thinking particular individuals morally important. For me and for Warnock the question of whether or not an individual is a person just is the question of whether it is morally important; and in particular of whether it shares whatever moral importance normal adult human beings have. I doubt anyone will be confused by this, but for our present purposes we do not need the term 'person' and I am content to re-pose Warnock's reformulation of my original question: when do embryos begin to matter morally?

Moral significance

The question then is, when if at all, and in virtue of what does the human embryo begin to matter morally? There are only two sorts of answer that might be given to this question. One is in terms of what the embryo *is*, that is in terms of some description of its morally relevant features. The other is in terms of what it will *become*, that is, in terms of its potential for acquiring morally relevant features. These two approaches seem to have a natural tendency to collapse one into the other as we shall see, but I shall start by trying to keep them distinct.

What is the embryo?

From this perspective the moral importance of the embryo might derive from different sorts of things that might be said about it. One strategy is to attempt to give an account of the sorts of features or capacities that might make for moral relevance and see which of them apply to the embryo at particular stages of its development. The other is to locate moral importance in one resounding central principle. In a famous essay on Leo Tolstoy, called *The Hedgehog and the Fox*, Isaiah Berlin[7] takes a fragment of Greek poetry and uses it to create a celebrated typology of human thought: 'The fox knows many things, but the hedgehog knows one big thing.' There are those, according to Berlin, who pursue many ideas and those who like to bring everything under one central vision or organizing principle. The latter are hedgehogs, the former are foxes.

1. The embryo is a hedgehog

The slogan 'human life begins at conception' nicely captures the hedgehog's approach to the moral status of the embryo. The hedgehog believes that what matters morally is being a member of the human species and that membership begins at conception.

Now in one sense this is simply false. The human egg is alive well before conception and indeed it undergoes a process of development without which conception would be impossible. The sperm too is alive. In one sense then life is properly seen not as beginning, but as a continuous process that proceeds uninterrupted from generation to generation. The genes too are alive of course, and many biologists speak of genes as 'immortal', in the sense that they continue from generation to generation, continuously copying themselves and expressing themselves anew in each generation.[8]

Well, if not life, nor yet human life, then at least the new human individual begins at conception. Again, this is at best a misleading claim. A number of things may begin at conception. Fertilization can result in the development not of an embryo but of a tumour, called a hydatidiform mole, which can threaten the mother's life.

Even when fertilization is on the right lines it does not result in an individual. The fertilised egg becomes a cell mass which eventually divides into two major components: the embryo and the trophoblast. The embryo or inner cell mass (epiblast) becomes the fetus and the trophoblast develops to form the fetal portion of the placenta and the umbilical cord. 'The trophoblastic derivatives are alive, are human, and have the same genetic composition as the foetus and are discarded at birth.'[9]

A further complication is that the fertilised egg cannot be considered a new individual because it may well become two individuals. This splitting to become 'twins' can happen as late as two weeks after conception.

The idea of human life having a beginning is clearly problematic. It is more plausible to regard it as a continuum with the human individual emerging gradually. If what matters morally is human life, we are faced with the problem of protecting all human life, including unfertilised human eggs and sperm. On the other hand if what is judged worthy of protection is not human life, nor yet living human tissue, but some fuller and richer description of just what it is that is morally significant about creatures like us, then we have to be more foxy.[10]

2. The embryo is a fox

The foxy approach to the problem of the moral status of the embryo is more sophisticated. Typically the fox will attempt to identify morally relevant features of the embryo or fetus and argue that in virtue of its possession of these characteristics or capacities it is worthy of protection. The fox's virtue is its ingenuity so there are many different accounts that might be given of the moral importance of the embryo. The following is not untypical.

> By eighteen weeks the unborn child is a fully formed, unique human being, with all its major organs, apart from the lungs, functioning. At this stage it is responsive to light, warmth, touch, sound and pain, and is even getting to know its mother's voice.[11]

The creature described in this passage, unique, fully formed with all its major organs functioning, and sentient—responsive to light, pain, and so on—could be the normal adult member of a thousand or more species of creatures that inhabit this particular planet. Many millions of such creatures appear on the dining tables of those who would protect human embryos, and millions more are used for experimental purposes. The only justifications that can be produced for distinguishing morally between such fully formed, unique, sentient creatures and the human embryo[12] are either the bare and bleak stipulation that members of the human species are morally important and indeed superior *per se*, or the view that the human embryo's moral importance lies not in what it is, but in what it has the potential to become.

Humans are the greatest

The belief that members of the human species are not only morally significant *per se*, but without a peradventure more morally important than any other creatures, is not without its attractions. Indeed we have good authority for its respectability. Mary Warnock makes it a central plank of her own position. Speaking of preference for the human species she says:

> Far from being arbitrary it is a supremely important moral principle. If someone did not prefer to save a human rather than a dog or a fly we would think him in need of justification . . . To live in a universe in which we were genuinely species indifferent would be impossible, or if not impossible, in the highest degree undesirable. I do not therefore regard a preference for humanity as 'arbitrary', nor do I see it as standing in need of further justification than that we ourselves are human.[13]

Repeating this thought more recently Mary Warnock has indicated that it was not only her own view but the unanimous view of the members of the Warnock Committee.[14]

Now it might plausibly be argued that to live in a universe in which we were genuinely species indifferent would be impossible or undesirable, in the same sense as it might be impossible to be genuinely, gender, race, religion, or nationality indifferent. But that does not mean that those with whom we do not share gender, race, or nationality are morally on a par with dogs or flies, or that we might defensibly so think of them. What makes racism or sexism wrong is not the simple fact that members of different races or genders are members of the same species but rather that there are no morally relevant differences between them.

Apart from its quite staggering complacency we should be warned against the assertion that species preference stands in need of no further justification 'than that we ourselves are human', if only because the same impenetrable preference has been asserted for race, gender and nationality with familiarly disastrous and unjustifiable consequences.

II. Potentiality

If the human embryo is not, in terms of what it is, significantly different from other creatures (including the Sunday roast) which we judge less morally significant, if in short it is not sentient, responsive, with organs functioning, etc.—if its only claim to a morally relevant difference is in terms of its membership of the human species—then we have no sufficient reason for according to it superior moral significance unless we can show some other morally relevant feature. The one remaining candidate of any plausibility is its potential.

If we assume for the moment that normal adult human beings, creatures like you and I, are morally significant if anything is, then it is the potential of the embryo to grow into such a creature that distinguishes it from all other creatures and their embryos, including those destined to constitute the Sunday roast. This argument has proved singularly attractive.[15]

There are two sorts of objections to the 'potentiality argument' for the moral significance of the embryo. The first is simply that the fact that an entity can undergo changes that will make it significantly different does not constitute a reason for treating it as if it had already undergone those changes. We are all potentially dead, but no one supposes that this fact constitutes a reason for treating us as if we were already dead.

The second objection is simply that if the potentiality argument suggests that we have to regard as morally significant anything which has the potential to become a fully fledged human being, and hence have some moral duty to protect and actualize all human potential, then we are in for a very exhausting time of it indeed. For it is not only the fertilised egg, the embryo, that is potentially a fully fledged adult. The egg and the sperm taken together but as yet un-united have the same potential as the fertilized egg. For something (or somethings) has the potential to become a fertilised egg, and whatever has the potential to become an embryo has whatever potential the embryo has.

Those who see in the potentiality argument the sole salvation of the embryo are quite naturally reluctant to accept the potentiality of the egg and sperm. They see a difference between the potential *of an individual* and the potential *to become an individual*. This distinction, if it identifies an important difference, makes it possible to regard the fertilised egg as a protected being without attaching any moral importance to the living human egg and the living human sperm which will become the embryo.

It is at this point that two sorts of defences of the embryo that we have thus far kept separate unite in their susceptibility to a powerful objection. For both defenders of a preference for the human species as such and those who value human individuals for their potential to become the sorts of beings that are morally significant—fully fledged adults perhaps—believe falsely that it is only the fertilised egg, the embryo, that qualifies. This is believed, so it would seem, because of acceptance of another false assumption, namely that it is only in the embryo that there is united in one place, in one individual, all that is necessary for continuous development to maturity.

This almost mystical reverence for the individual is never explained. Why is it right to protect individuals with the requisite potential but not pairs of individuals with the requisite potential?

However, another possibility also throws doubt upon the idea that the human individual begins at or following conception.

Parthenogenesis

The eggs of most species including humans can be stimulated to grow without fertilisation. This occurs naturally and randomly so far as we can ascertain, and may account for alleged examples of virgin birth, including that perhaps of Jesus; but only on the assumption that the son of the God of the Christians was in fact her daughter. For parthenogenesis only produces females. It is now possible to induce parthenogenetic growth of embryos grown *in vitro*. This possibility shows that the human egg is an individual member of the human species if the embryo is, for they both contain within the one individual all that is necessary for continuous growth to maturity under the right conditions.

This stipulation is important because there is some evidence that while parthenogenetically stimulated human eggs can develop normally until the stage when heart-beat is discernible, they will then usually die. As Martin Johnson has reported:

> eggs can be activated in the absence of spermatozoa and will proceed to develop as though fertilized, in many cases through to implantation and the development of a clear embryonic form with a beating heart. Only then do such 'parthenogenetically activated' embryos die and they do so because of the absence of the required activity in the placenta of the paternally imprinted chromosomes from the father.[16]

How far the inability of human embryos so far observed to grow beyond this early stage is a contingent matter, dependent simply on the development of appropriate means artificially to induce 'the required activity in the placenta', is difficult to say.

Recent research has shown[17] that the passage of cells giving rise to the gametes through either the female germ line in the ovary or the male germ line in the testis is not the same. It is believed that male or female germ cell formation brings a biological 'memory' to the mature gamete concerned so that the chromosomes 'know' whether they have male or female origin and the two are clearly different. This has repercussions for the embryo since the behaviour of the genes so 'imprinted' is different in the fertilized egg depending on whether they were derived from the male or female parent. The evidence apparently suggests that it is necessary to have genes imprinted in both the male and female manner to achieve a successful pregnancy. This means that the jury is still out on the question of whether successful human parthenogenesis is possible. It is of course possible that further research may allow the artificial imprinting of the chromosomes with the appropriate 'memory' or may allow us to by-pass this problem another way.[18]

For our purposes the fact that parthenogenesis can produce a 'clear embryonic form with a beating heart' indicates that it would be rash to assume that human parthenogenesis is not or will not become a possibility and that any moral theory which depends on this *not being a possibility* has made itself a hostage to fortune.

What is the point of potential?

There is one other point that should be made about the argument from potential. It is this, even if it were true that only the fertilised egg had the potential to become a fully fledged human being, we would have to face the question: why does this make it morally important? The answer must be because it is morally desirable to actualize human potential. That is, if the embryo or blastocyst is valuable only because of its potential then it must be the actualizing of this potential that is what is important from the moral point of view. But if this is so then we face another problem.

Those who accept that we ought not to kill or use or experiment on individuals with the potential to grow into morally significant beings have then a substantial problem as to what to do about all those women who wantonly insist on consigning to their deaths every month the human individuals which they could so easily take steps to nurture and protect.[19]

Warnock on potentiality

Mary Warnock herself is ambivalent on the question of potentiality. The Warnock Report took the view that 'the objection to using human embryos in research is that each one is a potential human being'.[20] In a more recent essay Mary Warnock herself takes the view that 'The question whether or not they may be used for research must be answered not with regard to their potential, but with regard to what they are . . .' Indeed, she goes further, confirming the arguments above: 'To say that eggs and sperm cannot by themselves become human, but only if bound together, does not seem to me to differentiate them from the early embryo which by itself will not become human either, but will die unless it is implanted.'[21] Now with this, of course, I entirely concur. However, Warnock continues with a gloss upon how we are to understand 'what they are', namely as 'how far they are along the road to becoming fully human'.[22] But unless they are morally significant in terms of what they are at the particular moment at which the judgement is made, that is at the chosen stage along the developmental path, then this simply collapses into the potentiality argument again. For if what they are is simply 'beings a certain way along the road to becoming something else, namely fully human', then their status derives from their potential not from their actuality.

Independent existence

There is a strong current of thought that attaches moral significance to the individual's ability to exist independently. For example, the

Infant Life Preservation Act 1929 makes it an offence to abort or end the life of an individual capable of being born alive. The convention has been to regard 28 weeks as the point at which a fetus is deemed capable of being born alive, and so abortions have 'traditionally' been permitted up to that point. However, advances in neonatal care have steadily reduced this time until the fetus may with luck and high technology survive at 22 weeks. This has in part lead to pressure for a new abortion act, and the Alton Bill was one recent attempt made in the British Parliament to reduce this cut-off point for permissible abortion to eighteen weeks. Indeed, many lawyers think that technology has simply overtaken the 'traditional' interpretation of the 1929 Act as permitting abortion up to 28 weeks, and that abortion is permissible under that Act only up to the point that technology cannot successfully aid the survival of a premature baby, at whatever stage of gestation this is at any given moment.

For the record, the Human Fertilization and Embryology Act 1990 reduces the permitted time limit for so called 'social abortions', that is abortions where the ground for termination is 'that the continuance of the pregnancy would involve risk, greater than if the pregnancy were terminated, of injury to the physical or mental health of the pregnant woman or any existing children of her family', to 24 weeks.[23] However the Act is more radical in other ways, permitting abortion to prevent fetal handicap up to birth. The Act also permits, with substantial reservations, the use of embryos for research up to the appearance of the primitive streak or up to fourteen days whichever is the earlier.[24]

To return to the issue at hand, the ability to exist independently is perplexing as a criterion. For one thing it makes the moral significance of the fetus turn on where it happens to be at the crucial moment. If its mother is near a major medical centre equipped to cope with premature babies, it might have a chance at 23 or 24 weeks. The further it happens to be from the premature baby unit the less its moral importance until the wilderness swallows it up entirely. This approach makes moral importance not only geography relative but technology dependent. We may be on the verge of an era in which some embryos will be fertilized *in vitro* and will grow to maturity entirely independent of their genetic mother, nurtured in an artificial womb—so called *ectogenesis*. Now this might be thought of as a form of dependence comparable to that on the mother; but only at the cost of regarding the equipment used to sustain premature babies in the same light, and hence regarding babies in incubators as lacking the requisite independence. Equally, all the technology from pacemakers to dialysis which sustain the life of so many adults would sustain them, at the cost of taking away their moral status and hence the arguments for supporting them at such cost.

The moral importance of the embryo cannot turn on something as slippery and as devoid of moral content as 'relative dependence'. Of course those who remain attached to the viability criterion may see in ectogenesis an argument for always preserving such independent embryos or fetuses. Here the argument is analogous to that provided by possible existence of the Sisters of the Embryo, whom we will shortly be meeting.

Sentimental morality

One suggestion that has been influential is the idea that moral sentiments must play a crucial role in the determination of what is morally permissible. This idea, originating with David Hume, has been influential in the work of a number of contemporary moral philosophers.[25] In particular, Mary Warnock has made it a central part of her own approach to these issues. It generates a strong and a weak thesis and we must be clear about both of these possibilities.

Briefly the idea is:

> If morality is to exist at all, either privately or publicly, there must be some things which, regardless of consequences should not be done, some barriers which should not be passed.
>
> What marks out these barriers is often a sense of outrage, if something is done; a feeling that to permit some practice would be indecent or part of the collapse of civilisation.[26]

As I say, this idea generates a strong and a weak thesis. The strong thesis is that where people's moral sentiments are outraged at the very idea of something, this fact of itself shows that what outrages them contravenes morality. The weak thesis on the other hand is that wise legislators will regard public sentiment as *strong evidence* about the morality of those expressing the sentiment and will therefore try not to violate or disregard these moral beliefs.

Mary Warnock herself seems to vacillate between these two, sometimes seeming to support the strong thesis and at other times the weak. For instance in her introduction to the Blackwell edition of the Warnock Report,[27] she adopts the strong thesis, using it as a form of veto. She argues that *only if* the advantages of embryo experimentation in utilitarian terms 'seemed very great' and *only if* 'there were no absolute outrage of general moral sentiment', should the embryo be used for research. However, in her more recent essay she suggests that the role sentiment plays is not so much that of a sort of veto or test that actions or practices must pass if they are to be permissible, but more an exercise in political compromise 'of attempting to come up with a moral solution to problems which, while retaining as many of the calculated benefits as possible, will nevertheless offend and horrify people as little as possible'.

Both the strong and the weak forms of this thesis suffer from the same fundamental weakness, that they both assume a sense of outrage is always a sense of moral outrage. Although Mary Warnock often talks of 'moral outrage' rather than simple 'outrage', she nowhere gives us an account of how to tell the one from t'other. While I myself think that David Hume's remark, quoted with approval by Warnock, that morality is 'more properly felt than judg'd of' is just plain wrong, and indeed that the converse is nearer the truth, even those who are inclined to accept it must face the problem of how to determine when one's feelings are moral feelings and when they are not.

This is a point of quite fundamental importance. For we can recognize that we ought, if we can, to respect people's moral beliefs or feelings even where we disagree with them (I say 'if we can', because we might not be able to do so without compromising our own moral beliefs). We can see that it is this belief, that we ought to respect the morality of others, which makes the weak thesis very attractive indeed, for it is one formal way of expressing our respect for morality itself.

But the crucial problem, entirely ignored by Warnock, is that not all feelings are moral feelings and not all outrage is moral outrage. So that while we ought to respect the moral beliefs and feelings of others even where we do not share them, we have no reason to respect their prejudices or brute preferences or aversions. Not only are we under no obligation to respect such things—they are not respectable. If we look at Mary Warnock's own elaboration of her idea we can see clearly why this must be so.

> Someone who feels that, for example, to shovel the dead into the ground without ceremony is wrong, may be able to say no more than that he regards such practices as unfitting or unseemly or uncivilized. But these very sentiments give rise to imperatives: one must treat the dead with respect. Similarly, ... those who object ... to commercial agencies for the supply of surrogate mothers may feel simply that they would be ashamed to live in a society where such agencies were permitted. To have such a feeling of shame must lie at the root of any moral principle.[28]

Now it is true, as we have already observed, that Mary Warnock constantly uses the prefix 'moral' when describing such feelings, but saying so does not make it so, and we are entitled to ask: just how does a feeling about something of moral importance become a moral feeling? It is of moral importance that black people are not unfairly discriminated against, but if I feel that they should be discriminated against my feeling is not automatically a moral feeling because it is about something of undisputed moral importance.

Feelings *per se*, of course, may well lie at the root of morality itself and it is not impossible that brute feelings lie at the root of any moral principle. But such feelings also lie at the root of many immoral principles and bare prejudices. We know of so many analogous feelings: that women are innately inferior and it is unseemly and unfitting for them to indulge in many 'male' occupations or to appear in public unless swathed from head to toe, that many people have felt ashamed to live in a society that permitted marriage between members of different 'races', or in which public institutions and places of resort were not racially segregated, or in which homosexuality was treated other than as 'a vice so abominable that its mere presence is an offence'.[29]

The first duty of someone who thinks that morality matters is to examine her feelings to attempt to see how and to what extent they cohere with her principles, whether they are simple personal aversions interesting only to the extent to which we are interested in the biography of the person whose feelings they are. And the first duty of those who are trying to determine public policy is to try to distinguish between expressions of outraged prejudice and expressions of outraged moral feelings. For even the weak thesis does not require us to respect bigotry, however passionately felt.

The moral sentimentalist[30] owes us an account of how to identify moral feelings and distinguish them from prejudices.[31] For unless she provides this, deference to moral feelings is indistinguishable from deference to prejudice, and the moral imperatives generated by sentimental morality are in effect injunctions to ignore moral reasoning altogether.

In order to know what bearing public sentiment is to have on the question of the moral status of the embryo, we need to know something about the grounds of the beliefs of which the sentiment is the expression. Now of course if nothing turns on the outcome, then there is no reason not to give way even to prejudice. If everyone feels, for whatever reason, that the dead should always be buried with ceremony, then if there is no reason not to do so this should certainly be permitted.

The problem, however, is that seldom does nothing turn on whether or not we respect such feelings. We can illustrate the problem with an example taken from another contemporary issue. Suppose we were to say that Salman Rushdie's book *The Satanic Verses*, to which the Muslim community throughout the world has taken violent exception, and which consequently has allegedly caused outrage to the *moral* sentiments of large sections of any community with a significant Muslim population, should be banned. This suggestion would have merit only on the assumption that freedom of speech is a trivial value in comparison with the moral outrage of a section of the community. But of course freedom of speech is not a trivial value. If we are faced simply with a clash between incompatible moral values, then we must find some way of choosing between them or of deciding which is more important. However, if one of the rival principles has doubtful claim to be a moral principle at all, then the reason to give it the same weight as a rival principle which can lay plausible claim to be a moral principle disappears. For the claim that we must respect another's morality is itself a moral principle, whereas the claim that we should respect another's prejudices is not.[32]

I cannot here attempt a positive account of what makes a principle a moral principle. However, at a bare minimum I would want to argue that someone holding what she claims to be a moral principle must be prepared to justify that principle in moral terms. That is in terms which would refer to the way in which violating the principle causes harm to persons or otherwise adversely affects persons or their interests or violates their rights or causes injustice.

In the case of *The Satanic Verses*, Muslims can certainly claim that they are caused mental anguish by seeing their religion, as they perceive it, insulted. However, while this is certainly a real harm it is one that, if recognised, would completely nullify a value like freedom of expression. For sensitive people do suffer harm when their sensitivities are trampled on, and if we are morally obliged to forbear whenever this is the case then freedom of speech would be 'hostage' to the most sensitive (whether rationally sensitive or not) individual or group in the community.

Moreover, it is a cliché that nothing is so hurtful as the truth. If we are entitled to be defended from learning this (or what others believe this to be) about ourselves or things we value, then the ability to mount a critique of any existing values, beliefs, or prejudices will always be contingent on how strongly these beliefs or values are held and consequently how much pain is involved in seeing and hearing them challenged.

It is time to remind ourselves of what moral reasons we might have for permitting the use of embryos for research and as sources of cells and tissue.

Why experiment on embryos?

There can be few people who do not know the role that research using human embryos has played in the treatment of infertility and in evolution of the techniques that have made so called 'test-tube' babies possible and which have helped so many couples to have the children that they so passionately want. Less well publicized are the possibilities that are opening up of using embryo research for diagnostic and therapeutic purposes.

In vitro fertilization offers the possibility of identifying genetic abnormalities in the embryo before implanting it into the uterus.[33]

The same techniques permit the determination of the gender of each embryo with the consequent possibility of screening effectively for sex-linked genetic disorders. There are also very good indications that embryo or fetal cells, tissue, and organs can be used for repair and transplants in adults. This may make possible the repair of inherited enzyme defects, the treatment of diabetes using pancreatic cells, and embryonic myocardial tissue could be obtained from embryos and used by cardiologists to repair the major vessels of the heart.[34]

Robert Edwards reports[35] that work is far advanced on mouse embryos which if repeated in the human embryo will provide a good chance of repairing lethal radiation damage. It has recently been reported elsewhere[36] that tissue from the brain of an aborted human fetus of around thirteen weeks' gestation will be used in the treatment of Parkinson's disease.

While it is not clear precisely which lines of research will be most useful, nor what other promising possibilities will arise in the future, we do know that none of them are likely to be realized without embryo research. If, as seems overwhelmingly probable, embryos can be used to save the lives of adults and children and for therapeutic and diagnostic purposes, we would require strong moral arguments indeed to justify cutting ourselves off from these benefits with the consequent loss of life and perpetuation of pain and misery. Before returning to the question of whether such arguments are available we must pause to consider whether morally relevant distinctions can be made between embryos.

Spare embryos

There are two different sources for embryos (or indeed fetuses if the embryos are grown long enough) that might be used for research or therapeutic purposes. The first are those grown externally in the laboratory. These may not all have been conceived *in vitro*, for the process called lavage is a way of recovering from the mother, not eggs, but early embryos which have been fertilized *in vivo* not *in vitro*. However, all these are early embryos and they all end up growing *in vitro*, wherever they started out, so I shall for convenience call them 'in vitro' embryos. The second source are embryos or fetuses, usually at a much later stage of gestation, which are the products of abortion.

Now among *in vitro* embryos we need to make three further distinctions. There are those healthy embryos which have been grown with a view to implantation in a mother, with the hope that they will grow into normal adult human individuals. These may be 'fresh' growing in the laboratory or frozen for future implantation. The second group of embryos are those which are sometimes called 'spare' embryos. The hormone treatment which produces extra or 'superovulation' in women, so that a number of eggs can be recovered at a time, may lead to the existence of embryos which are spare in the sense of being more than are required for implantation either presently or in the future. These spare embryos may be normal or anomalous. Those with anomalies of any sort—possible defects which might prevent normal development—would not be implanted. Normal spare embryos are suitable for implantation but are superfluous to requirements. The common feature of all the embryos so far discussed are that they have been produced as part of a programme designed to result in embryos for implantation.

Finally there are what have sometimes been called 'research embryos'. These are not produced with a view to implantation but are gathered for research purposes only. The usual source for these embryos are women who wish to be sterilized and who are asked if they are prepared to donate eggs for research. These eggs are then fertilised and the resulting embryos constitute the so-called research embryos.

Robert Edwards[37] draws a firm distinction between using spare embryos and research embryos, arguing that there is something degrading about producing embryos at will merely for research purposes. He believes it permissible to use only those spare embryos that are anomalous in some way and that would consequently be unsuitable for implantation in any event. There are three possible interpretations of this unease. One focuses upon the moral character of the researchers, arguing that those who would deliberately choose to create embryos for their own purposes alone are somehow morally deficient. The second is a form of Warnockian argument, suggesting that the feeling that one would not wish to live in a society that permitted such things is of itself evidence of their immorality. And third there is an oblique reference to the potentiality argument, the suggestion being that anomalous embryos are not even potential human individuals and so there can be no harm in using them, whereas normal embryos should be implanted and given their chance to develop.

The first two interpretations are good reasons for Professor Edwards to refrain from doing things that make him uneasy. The question is though, whether others who do not feel as he does should refrain on these grounds? The only difference between using spare embryos and research embryos that we have not already considered is the intention of those who produce them. Spare embryos are produced in order to establish a successful pregnancy and research embryos are produced in order to do research or to provide sources of tissue or cells for beneficial use in others. There is no difference in the moral status of the respective embryos, nor is there any difference in what will actually happen to them. The difference is only in the initial intention behind their production. I cannot but think that if it is right to use embryos for research then it is right to produce them for research. And if it is not right to use them for research, then they should not be so used even if they are not deliberately created for the purpose.

If, as is believed, the 'cost' of producing a live birth by normal sexual intercourse is statistically the loss of between one to three embryos in early miscarriage or failure to implant, and it is not wrong knowingly to attempt to have children at such a cost, then, one might be forgiven for thinking that it is not wrong to attempt to save the lives of other human beings at the same or a comparable cost.[38]

Those who believe the ends do not justify the means and who consequently believe it would be wrong to sacrifice an embryo in order to have a live birth should of course never practice 'unprotected' sexual intercourse. For this will result, almost certainly, in their eventually having a live birth 'over the dead bodies' of up to three embryos that have aborted spontaneously and in all probability unobserved. If this is right it would seem that Catholics, so far from outlawing contraception, should rather outlaw attempts to conceive. For if they believe it would be wrong knowingly to sacrifice innocent life, and if the ends do not justify the means, and if the early embryo does constitute 'innocent life' in this sense, then they do knowingly and deliberately sacrifice innocent life when attempting to have children.

A Catholic response might be to insist that in such cases there is no intention to kill embryos and, using the doctrine of double effect, argue that one is only morally responsible for outcomes directly intended rather than for those which occur simply as a second but *ex hypothesi* unwanted effect of what is intended. I have argued, I believe decisively, against the moral relevance of the doctrine of double effect elsewhere;[39] however, for the moment we may observe that intention is a highly slippery notion and those who wish to experiment on embryos can claim with equal plausibility that their primary intention is to find therapies and if embryos must die as a second but unwanted effect of such a quest, then this must, by parity of reasoning, be permissible. The point is surely that we must take responsibility for what we knowingly and deliberately bring about, not simply for what we are hoping for.

Potentiality again

The idea that there is a moral gulf between embryos destined to be implanted and embryos not so destined is tenacious. It is true that there is often a moral difference here but it is not a difference in the moral status of the various embryos, but rather a difference in the account we give of the wrong that might be done if they are damaged in any way. In the case of an embryo destined for implantation, to harm that embryo would be to wrong the woman (and perhaps her partner) who wishes to receive that embryo. It would not be a wrong to the embryo so long as the embryo is not in fact implanted.[40]

Of course if it is damaged and then implanted and it grows up into a damaged person that person will have been wronged or harmed by the harm done to the embryo. We will be returning to this complex but interesting question in Chapter 4.

There is something essentially slippery about according different moral status to embryos on the basis of their supposed destiny. To see this we must examine the effect on the argument of the existence of a mysterious and fictitious religious order.

Sisters of the Embryo

The 'Sisters of the Embryo' are a useful but fictitious religious order of women whose one purpose is to provide homes for homeless embryos. They stand ready to offer their uteruses to any and every embryo, whether normal or anomalous, fresh or frozen. So long as an order such as the Sisters of the Embryo is possible, then no embryo is truly spare or homeless or lacking in potential. After all, every embryo without an immediate prospect of implantation could be cryopreserved until a uterus could be found for it (that is until the Sisters of the Embryo come into being). So even the hypothetical existence of the Sisters of the Embryo shows us that, whether or not a particular embryo has a uterus waiting for it, all are in the same moral boat. For even an anomalous embryo might want, or at least have an interest in, the chance of an anomalous life.[41]

The moral status of the embryo

We must return finally to the question of the moral status of the embryo. We have seen that this cannot be determined by its potential, by what the embryo will become, but must be assessed in terms of what it is. I have set out in detail elsewhere my own arguments about the criteria for moral significance of the individual, and my answer to the question of when and in virtue of what embryos might begin to matter morally.[42] I argue that the moral status of the embryo and indeed of any individual is determined by its possession of those features which make normal adult human individuals morally more important than sheep or goats or embryos. Now, while I have a positive account of what these features are, I do not need to deploy it here. For it is clear that at no stage of its development does the human embryo nor yet the human fetus possess these characteristics. This was clear in our consideration of 'foxy' arguments about the status of the embryo.

The crucial issue seems to be this: we have no reason to think that the embryo, nor yet the fetus, attains a moral status comparable to that of adults at any stage of its three trimesters of gestation, whether these occur in the womb or not. The changes that occur in the developing embryo while real and fascinating do not make for important moral differences. At no stage does the embryo or the fetus become a creature which possesses capacities or characteristics different in any morally significant way from other animals. It differs from other creatures to be sure in its membership of the human species and in its potential for development to human maturity. But in these respects it does not differ from the unfertilised egg and the sperm or, if parthenogenesis is a real possibility, from the unfertilized egg alone.

When we bear in mind that, as Robert Edwards has argued,[43] most of the secrets of the development of life are contained in early embryos, and that we are extremely likely to be able to use what we learn from such embryos to save many lives and ameliorate many conditions which make life miserable, we would not only be crazy but wicked to cut ourselves off from these benefits unless there are the most compelling of moral reasons so to do. I have argued that there are no such compelling reasons.

3

Origins and Terminuses

'Life is so terrible; it would be better never to have been conceived.'

'Yes, but who is so fortunate? Not one in a thousand.'

This Jewish joke, first given philosophical prominence by Robert Nozick, in a footnote to *Anarchy, State and Utopia*,[1] draws attention to the feature that is so often adduced in justification of pre-natal screening and indeed of abortion, namely that life can be terrible and that this fact provides grounds for preventing a terrible life from beginning or continuing. The conditions that make it so can be intrinsic or extrinsic. The child can be disabled or damaged or injured in some way that makes life less than optimal, or the circumstances into which a healthy child is born may occasion the unsatisfactory nature of its existence.

The ethical problems surrounding pre-natal screening and termination of pregnancy can then be approached from two very different perspectives. One of these involves establishing the moral status of the human embryo, which is a necessary precondition to understanding what one might be entitled to do to, or with, it. This dimension we have just examined.[2]

The other perspective involves examining the question, not of whether or not we might be entitled to kill the fetus or indeed experiment on it, but rather of whether or not we are sometimes morally justified in bringing it into existence or in allowing it to continue in existence.[3] This is a vital question both for its own sake and because when considering the ethics of bioengineering, we will want to know when we are entitled to bring into existence genetically engineered or otherwise modified individuals. We will also want to know how far we are entitled to use genetic screening as a method of choosing between possible rival candidates for existence.

This general problem has also two distinct dimensions. The first involves an examination of potential children for their adequacy as children; and the second involves examining potential parents for their adequacy as parents. Or, to put the point another way, one dimension of the problem involves asking whether we might do wrong by bringing particular children into existence because of problems relating to what one might call the constitution of those children, in virtue of which we might expect them to have less than adequate or satisfactory lives. The second concerns the question of whether we might do wrong by permitting children to be brought into existence who will suffer from less than adequate parenting.

So the ethics of pre-natal screening must be understood to include both pre-natal screening of potential children and of potential parents.

The pre-natal screening of embryos is becoming more and more common and more and more effective. We can now screen for all sorts of disorders from Down's syndrome and spina bifida to Huntingdon's chorea and AIDS. This process will often result in termination of pregnancy where screening gives warning that a child will, or will probably, be handicapped or be at substantial risk of developing a disorder. In this chapter I will say something about whether and why such practices are ethical if they are. Certainly these practices presuppose a view about what is a reasonable quality of life for a child to be born with, or to expect, or a view about what quality of life it is reasonable to inflict on a child when we can avoid so doing by terminating pregnancy.

But it is not only embryos or indeed potential embryos, gametes, that are screened. As I have suggested, there is another and equally important dimension of pre-natal screening—it involves very often the screening of parents.

The recent House of Lords decision on sterilization is a case in point.[4] In this case, Jeanette, who was described as having a mental age of about 5, was sterilized at the instance of her mother on the grounds (among others) that she was unsuitable as a parent. Now, this decision and other recent cases in which mentally handicapped young girls have been sterilised or have had abortions ordered by the courts have all involved judgements about the fitness of those girls to be parents.

Judgements to the same effect (though of course far from similar) are made when candidates for example for *in vitro* fertilization are turned away on grounds that take into consideration concern about their quality as potential parents.

These and other cases challenge the supposed right to found a family, a right which is protected by Article 12 of the European Convention on Human Rights and by Article 16 of the Universal Declaration of Human Rights. I will be returning to these issues later.

Our question then is: are there any constraints upon who should be allowed to be a child and who should be allowed to be a parent? In what follows I will try to say something in answer to both these questions. In doing so I will comment upon the recent proliferation of legal cases in the United States in which children have brought actions against their parents or parents against doctors alleging so-called 'wrongful birth' or 'wrongful life'. The alleged wrong is, as we shall see, that a child has been allowed to come into existence in less than optimal circumstances. I will also have something to say about just when exactly the human individual may first be said to come into existence.

I shall begin by considering the question of whether or not we can do wrong by bringing someone into the world and say something about the underlying principles that might justify pre-natal screening and one of its consequences—namely the abortion of embryos found to be defective in some way. I shall then briefly consider the question of screening parents for their suitability as parents and in the next chapter I will discuss wrongful life.

I. Can We Do Wrong by Bringing Children into Being?

It is usually assumed in discussions of abortion that the only moral issue is whether or not it is justifiable to abort a viable fetus—whether that justification is seen in terms of women's rights, or in terms of protecting the health of the mother, or in terms of what is sometimes called the moral status of the fetus. It is not usually thought that the person who decides not to have an abortion, who decides to go ahead and have a child, might be doing something culpable.

One reason for this is the great attraction of having children. Without commenting on the rationality or indeed on any other element in the cogency of such a desire, there can be no doubt that having children is almost universally acknowledged to be one of the most worthwhile experiences and important benefits of life. This is perhaps why people do not usually think that there might be any necessity to justify their decision to have children or to explain why they have chosen not to have an abortion. This is also perhaps why it is the case that if someone wants to have a child, we usually think that they ought to have every assistance; and perhaps also why people have sometimes thought in terms of the right to found a family. But might we do wrong in founding a family? Let's pursue this a bit further.

Mill's argument

A classic formulation of the strongest argument against the freedom to reproduce is perhaps still that of John Stuart Mill in his essay *On Liberty*.[5]

> It still remains unrecognised that to bring a child into existence without a fair prospect of being able, not only to provide food for its body but instruction for its mind, is a moral crime, both against the unfortunate offspring and against society.

Now of course the crimes to which Mill refers are relatively easily remedied, as Mill is himself aware, for he continues:

> if the parent does not fulfil this obligation, the State ought to see that it is fulfilled, at the charge, as far as possible, of the parent.

So that we need not prevent people from reproducing but merely ensure that their children are properly educated and cared for. However, the powerful argument that Mill introduces involves the idea that one can harm people by bringing them into existence under adverse conditions. And this leads us on to the question of just how, precisely, we are to characterize the moral crime that we might commit by bringing children into the world in less than favourable circumstances.

What is the wrong that we do?

To consider this question we will have to do an experiment. Not perhaps the sort of experiment that one might expect from medical science. This involves an experimental technique familiar in philosophy since at least the time of Plato. It is a thought experiment. This technique involves inventing examples, sometimes annoyingly fantastical, in which we can so manage the variables as to enable us to concentrate on just the question at issue. It is the problem of managing the variables that sometimes makes the examples seem so fantastical—so-called 'desert island' examples—but our present thought experiment is fantastical only in the sense that the examples presuppose unknown medical conditions and undeveloped medical science. They derive from the philosopher Derek Parfit.

Parfit's argument

Parfit invites us to consider the actions of two different women:

> The first is one month pregnant and is told by her doctor that, unless she takes a simple treatment, the child she is carrying will develop a certain handicap. We suppose again that life with this handicap would probably be worth living, but less so than normal life. It would obviously be wrong for the mother not to take the treatment, for this will handicap her child . . .
>
> We next suppose that there is a second woman, who is about to stop taking contraceptive pills so that she can have another child. She is told that she has a temporary condition such that any child she conceives now will have the same handicap; but that if she waits three months she will then conceive a normal child. And it seems (at least to me) clear that this would be just as wrong as it would be for the first woman to deliberately handicap her child.[6]

Now, we can describe the actions of the first woman as acting in the best interests of her child, what she does prevents that child from becoming handicapped. To fail to take the treatment would be deliberately to handicap her child, to take the treatment is in her child's best interests.

However, the same seems not to be true of the second woman. For either there is no child whom she benefits by her action, because, when she takes the treatment, no child exists that she can thereby benefit, or it is in the interests of the child that she can conceive this month that she does so and that it thereby comes into being. If she postpones pregnancy and takes the treatment she damages the interests of this child for she removes its only chance of existence.

There is a considerable and important philosophical problem as to how precisely to characterize what is going on here. For clearly the second woman can and does benefit her child by taking the treatment.

The first alternative—that because no child exists at the time she takes the treatment the mother does nothing that benefits her subsequent child—is false. Both women can have healthy children as a result of their decisions, both women take steps which result in their having healthy children, two healthy children consequently exist. The actions of the two women are therefore morally equivalent and both have children each of whom have benefited from their mothers' decisions.[7]

The second argument raises more problematic issues. The suggestion is that the second woman injures (or damages the interests of, or acts maleficently towards) her child by postponing conception and pregnancy and thereby causing the child not to exist. But, since the child is as yet unconceived it does not exist anyway. Moreover and more importantly, it never will exist, and for both these reasons cannot be harmed. Therefore, like the first woman, this mother acts in the best interests of her child—the child that will exist as a consequence of the actions she takes.

However, this explanation seems to me to be suspect. As Derek Parfit persuasively argues,[8] 'Unlike never existing, starting to exist and ceasing to exist both happen to actual people. This is why we can claim that they can be either good or bad for these people.' This raises the question of when someone can properly be said to start to exist. Thomas Nagel, whom Parfit quotes with approval says: 'All of us . . . are fortunate to have been born. But . . . it cannot be said that not to have been is a misfortune.'[9] However, the second set of omission points in this quotation provided by Parfit are significant for they indicate a missing proviso. Nagel added the qualification that 'unless good or ill can be assigned to an embryo or even to an unconnected pair of gametes it cannot be said that not to be born is a misfortune'.[10]

Now Nagel clearly thinks that embryos and unconnected gametes do not mark the start of a person.[11] If we take a rich conception of what it is to be a person this is of course true. By 'rich' I mean a conception which does not merely see persons as co-extensive with humans but rather attempts to understand what makes for personhood, what makes persons distinct sorts of individuals. On such a conception it will also be implausible to think of the person starting at birth.[12] But, in the sense in which it might begin to make sense to talk of one's actions benefiting a particular person, it is surely implausible not to think of the egg which eventually becomes a person as the start of that person.

There are now over 1,000 babies, who will or will have become persons, who will have cause to thank Robert Edwards and Patrick Steptoe for their beneficial interventions. The gathering of the particular egg and its fertilization *in vitro* marks the start of the life of Louise Brown and the things that were done prior to conception, to the egg that became Louise Brown, were actions which she has good cause to regard as having been beneficial to her. So that while Nagel was correct to be doubtful as to whether good or ill can be assigned to an embryo or even to an unconnected pair of gametes, good or ill can be done to the person those gametes will become and that good or evil is done while that person is at the gametes stage.

Starting to be a person

There is an important theoretical point at issue here, which also, as we shall see, may have some practical importance. To explain this point we must go back to Nagel. The point of saying that 'All of us . . . are fortunate to have been born. But . . . it cannot be said that not to be born is a misfortune'[13] is simply that while those of us who were born exist to be fortunate and to benefit from coming into existence, those who are not born do not exist to suffer the misfortune of non-existence. So that to cause someone to exist is to benefit that person, but to cause someone not to exist by failing to bring them into existence, harms no one; for the simple and sufficient reason that there is no one who suffers this misfortune. Again as Parfit says:

> When we claim that it was good for someone that he was caused to exist, we do not imply that, if he had not been caused to exist, this would have been bad for him. And our claims apply only to people who are or would be actual. We make no claims about people who would always remain merely possible. We are not claiming that it is bad for possible people if they do not become actual.[14]

But of course once someone has started to exist, then it might be bad for them to be caused not to exist. I say 'might' be, for to understand what this might mean we need to say more about the idea of something's being bad for someone, and also more about the nature of the someone that it is bad for.

It seems sensible to say something like this: either a person starts to exist when she starts to be capable of benefiting from things done or not done to her (I do not beg the question as to whether an individual's life literally begins at all, or rather whether, as I think, it is more correct to say that life is a continuous process with the individual gradually emerging) or a person starts to exist at the point at which the developing [human in this case] individual becomes a person in some rich conception of the person. It seems that persons are capable of benefiting from things done or not done to them at the gametes stage and so it seems appropriate to say that an individual's life story begins, she starts to exist, when the gametes from which she will develop are first formed.

This view is certainly not widely accepted as we shall see and it is important to explore further the reasons for holding it.

It looks very much as though Parfit is committed to the first conception of the start of an individual's existence, although he is ambivalent to say the least as to when to date the start. Immediately following his approving quotation from Nagel which seems to date the start of life at birth, Parfit continues:

> . . . if it benefited me to have had my life saved just after it started, I am not forced to deny that it benefited me to have had it started. From my present point of view there is no deep distinction between these two. (It might be denied that it benefited me to have had my life saved. But, if this is claimed, it becomes irrelevant whether causing someone to exist can benefit this person. I ought to save a drowning child's life. If I do not thereby benefit this child, this part of morality cannot be explained in person affecting terms.[15]

It seems clear that Parfit believes that it does benefit me to have my life saved immediately after it started and he insists, rightly it seems to me, that if this is so, then it would be strange to say that it did not benefit me to have had it started. And this is consistent with my denying that it would have harmed me had my life not started at all—for there would then have been no-one who suffered this harm. But on this view one should date the start of *a life* (as opposed to *life*) at the first point at which one can identify an individual whose life can be saved. Again the point of distinguishing between life and a life is to leave open the question as to whether or not life is continuous.

To return for a moment to Louise Brown, if the egg that became Louise Brown would have died had it not been 'harvested', then Louise Brown benefited, in person, from that harvesting and from the subsequent fertilisation *in vitro*. In person-affecting terms, Louise Brown was favourably affected by those acts. She was favourably affected both because in a real, though not unproblematic, sense they were acts done to her, and because Louise Brown is the person who developed from that egg and who benefited from its fertilization and survival.

We should not forget the sperm. The reason for failing to take much interest in the male contribution to all this is that with present technology we do not usually identify in advance of conception the particular individual sperm which contributes so much to the proceedings.[16] However, there are many reasons to take more interest in the gametes from both sides of the equation. When we do identify individual sperm in advance of conception and perhaps even select among sperm for particular features we want, or indeed engineer those features into particular sperm, then resulting persons will be perhaps more conscious of tracing their origins equally to egg and sperm and will think of themselves as benefiting in person from what has happened to either of the sets of gametes to which they owe their existence.

Now, if this seems strained, we have to identify the strain. One possibility is the strangeness of calling an unfertilized human egg in a laboratory dish 'Louise Brown'. Of course we identify a particular egg as 'Louise Brown' with the benefit of hindsight. But so far from being a disadvantage, it is notorious that hindsight is the most accurate vision of all—twenty/twenty every time. But the point is this. If Louise Brown did not start to exist as an egg, then when did she start to exist? This is not the place to rehearse all the possibilities. Birth is clearly a non-starter. Nagel talks about a brief period of premature labour as being the only period before birth that one could sensibly claim the same person to have started to exist, but, unusually for Nagel, this is a hopelessly primitive view. There can be no significance in birth as the start of the individual person—her life can have been saved a thousand times prior to birth and we are on the verge of an era in which individuals will grow to maturity without ever having been 'of woman born'. If we regress from birth, through the development of the embryonic brain, past the formation of the primitive streak and past conception towards the identification of the individual egg, we cannot avoid the problems of morally significant person-affecting actions and omissions at every stage.

Of course there is a sense in which someone who saved the life of Louise Brown's grandmother benefited Louise Brown; but these person-affecting actions did not *happen to* Louise Brown although she benefited from them.

One alternative is to try to identify criteria for personhood in a rich sense, and argue that person-affecting only begins when we have a person properly so called.[17] The problem here is that the person obviously benefits from things done and omitted to be done to what I shall call the pre-person. I use the term 'pre-person' to make clear that I am in no way seduced by the 'potentiality argument', the argument that suggests that moral importance is conferred on an individual because of that individual's potential for development.

Harming and benefiting actual people

If we distinguish between human individuals who are potential or pre-persons and human individuals who are actual persons then we may find a defensible asymmetry between the various benefits and harms that may befall such individuals. Failures to benefit a potential or pre-person, or harms done to such an individual which result in her death, are harms to that pre-person but are not harms to the person she might have become because that person does not exist at the time the harm is done and will in fact never exist. In this respect causing the death of a potential or pre-person is morally on a par with failing to bring a person into existence. Whereas benefits done to the pre-person which save her life are benefits to the actual person she will become if and when that person starts to exist.

This is an approach which solves a number of problems and at the same time can be defended by at least one plausible approach to the question of what constitutes a person properly so called. First, it solves the problem which so concerned Parfit, that of how to explain why causing someone to start to exist benefited them while failing so to do did not harm them. And of course it solves that problem in the same way that Parfit solved it. But it takes Parfit's account further by showing when it may properly be said that an individual's life begins. Secondly, it performs the further task which Parfit's account looked as though it could not perform. That is the task of explaining our different reactions to abortion and perhaps also to infanticide on the one hand, and to murder on the other.[18] It also allows us to explain the moral symmetry between the actions of the two women considered earlier and indeed permits the drawing of distinctions and conclusions that are to come.

When an individual's life begins

The important difference between the present argument and Parfit's is that we have identified the point at which a particular life may be said to begin, namely when the egg from which it grows is first differentiated or when the sperm that will fertilize that egg is first formed, whichever is the earlier. In short:

> an individual's life begins when one of the gametes from which that individual will develop is first formed.

In most usual cases it will be the egg that is differentiated first, long before the sperm that will fertilize that egg is formed. However, the existence of techniques of cryopreservation, or more popularly 'freezing' both sperm and eggs, make sure that this is not necessarily the case. This is important because as Parfit has said, 'if it benefited me to have had my life saved just after it started, I am not forced to deny that it benefited me to have had it started'.[19] But surely it appears that if it benefited me to have my life started then it would have been bad for me to have been caused not to exist *after* my life started. So unlike missing out on being caused to exist, having one's life ended after it has begun, by abortion for example, looks as though it ought to be bad for that individual.

When Parfit says 'Causing someone to exist is a special case because the alternative would not have been worse for this person'[20] he seems to imply that this is the only special case of this sort because starting a life is a singular event. And because Parfit also believes that 'it benefited me to have had my life saved just after it started', he implies that to fail to save a life just after it started, or to kill it then, is bad for that individual. 'I ought to save a drowning child's life. If I do not thereby benefit this child, this part of morality cannot be explained in person-affecting terms.'[21] However, on the present view the asymmetry continues until the individual becomes a person at some significant time after birth. So that we say that an individual's life begins when the egg is first differentiated and maybe also when the sperm that will fertilize that egg is first formed. If that individual becomes a person that person will have benefited in person from the fact that the egg and the sperm started to exist and from all the person-affecting things that happened to both before fertilisation and to the embryo, fetus, and so on thereafter.

So that starting the life of a person and benefiting that person by things done to it as a pre-person are good for that person but to end its life or to fail to save it at any time prior to its becoming a person would not have been worse for this person; for the same reason that it would not have been worse for him if he had not started to exist.

If I am wrong about this and Parfit's account is sound, then, other things being equal, we do wrong not to fertilise any egg capable of fertilization, unless of course a more plausible event can be identified as the start of a life.

These have been complex points and I will briefly sum up this part of the argument. Parfit did not really address the problem that concerns us here. For this reason perhaps, he has left an account which is puzzling on the question of the morality of the treatment of pre-persons. His account seems to impel us towards dating the start of an individual's life as the first point at which she can benefit in person from things done or not done to her. Parfit seems, on crucial occasions, not to question Nagel's suggestion that this point is at birth or immediately before birth, although at other places he seems to accept that an individual's life starts at the point at which the unique genetic make-up of the emerging individual is first present. That is, when the zygote is first formed.[22] But Parfit also persuasively suggests that the moral obligation to do things to or to refrain from doing things to individuals must be justified in person-affecting terms. That is, their morality turns in large part on the effect they have on persons. Persons are affected by what happened to them as pre-persons. However, doing things to pre-persons which prevent them from ever becoming persons cannot be handled by a person-affecting morality, except in so far as they have side-effects which affect persons. This leaves us with a puzzling asymmetry between obligations to individuals who do become persons and obligations to individuals who do not. Parfit treats this asymmetry as if it is created by the singular event of starting to exist, but that thereafter the symmetry is restored. I have tried to show that this asymmetry must persist until the person has emerged from the pre-person.

The Zygotic Principle

One widely accepted suggestion for the point at which an individual's life can be said to begin, and one that is importantly at odds with the account developed here, is the formation of the zygote. This has been called 'The Zygotic Principle' and it is associated with the American philosopher Saul Kripke.[23] This is of course a rather pretentious way of describing a very well-established idea—namely the idea that life begins at conception. However, since I hope it is clear that I regard this view, however expressed, as erroneous, and since the term 'Zygotic Principle' is used in the work we are about to discuss I will stay with this term.[24]

In a recent essay Bernard Williams has defended the 'Zygotic Principle', which he dubbed 'ZP', and has explained the motivation behind it.

> The motivation of ZP lies rather at a more general, metaphysical, level in the importance of *origins* to our idea of a particular living thing (and of other kinds of things as well). When we speculate about possibilities, we consider different possible life-stories. We have to distinguish between *life-story of a different individual* and *different life-story of the same individual*. We understand the latter in terms of forward-branching alternatives in the life-story of the individual, with branches running from points at which what might have happened to him diverges from what actually happened. Since they provide alternative versions of one life-story, they are branches from one stem; and they are forward branching because the notion of a life-history is the notion of a causally structured sequence, a series in which later stages are explained by earlier stages. ZP provides the minimum basis for the stem of such a structure.[25]

Now the Zygotic Principle is certainly responsive to the motivation Williams identifies. However, it is not *fully responsive* to that motivation because, as we have seen, it is not only plausible but compelling to date for example the start of Louise Brown's life story from the harvesting of the egg from which she developed.

The Gametic Principle

I have suggested in this chapter that we should say that a persons life story begins when either of the gametes from which that individual will develop are first formed. If I were not embarrassed by the pretentious name I would call this: the 'Gametic Principle'.

Both principles have their problems. The problem with the Gametic Principle is that the gametes are not 'an individual' whereas the zygote is, or rather appears to be, an individual. This creates the problem of 'fusion', of whether when two individuals fuse to form one, the resulting individual can retain the 'identity' of the two with which it is continuous or whether we must say a new individual has then been formed. One problem with the Zygotic Principle is, as we have seen, that it does not allow the life story of an individual to begin when it is natural to say it begins. Another is the problem of monozygotic twinning, twinning where a single cell splits into identical twins with the same genetic make-up or where, at the blastocyst stage, the clump of cells that forms the blastocyst forms into two clumps which each develops as a separate identical twin embryo.[26] This, for obvious reasons, is often called 'fission'. It is twinning that makes the zygote's claim to be an individual problematic. As Bernard Williams concedes, the problem is that if the Zygotic Principle means that 'any individual who developed from that zygote would count as A, then . . . twinning is fatal to it: any story about A would equally be a story about A's sibling, and conversely'.[27] To overcome this supposed problem Williams proposes a modified Zygotic Principle.

The modified Zygotic Principle

Williams's paper was part of a symposium and was followed by discussion. During the discussion the embryologist Anne McLaren put the twinning objection to Williams and under pressure from her this was his response:[28]

> *a story is about A if it is about an individual who developed from the earliest item from which A in fact uniquely developed*, where 'uniquely developed' refers to development in the course of which there is no splitting. Where there is no twinning, this is equivalent to the original principle—the 'earliest item' from which A uniquely developed is the zygote. Where there is twinning, the history of A goes back to the twinning and no further.[29]

Williams wants to insist that the story never goes back further than the point at which A *uniquely* developed from something. Williams seems to have two different things in mind when he talks of *unique* development. On the one hand he seems to imply uniqueness in the sense of the individual not requiring any further genetic contribution, so that the individual's genetic nature is unique and complete at that point. Another idea is clearly that the 'earliest item from which A uniquely developed', must imply that A and only A developed from that item.

It seems to me that both these requirements put unnecessary strain upon the Zygotic Principle. If we remember Williams's, surely correct, explanation of the motive behind the principle, or as I would prefer to say 'its point', we can see why this is so. Recall that Williams rooted the point of the principle in the metaphysical idea of the importance of origins in our conception of just what something is. In particular that 'the notion of a life history is the notion of a causally structured sequence, a series in which later stages are explained by earlier stages'. Now surely in both senses an individual's origins are rooted in both gametes,[30] particularly where what happens to those gametes is an important part of the life story of that individual. We have already concentrated on one case, that of Louise Brown, but somewhat further back in our discussion we considered cases in which an individual's germ line or somatic line might be altered at the gametes stage, giving that individual immunity to genetic disorders or even to diseases like AIDS or hepatitis B. Or the individual might have a new gene inserted at the gametes stage thus giving the resulting individual an entirely new and unprecedented genetic structure, a new, and possibly not quite human, nature.[31] What could be more important to an individual's identity than such an event? This would surely be considered by someone as not only an important event in their life history but also an important element in whatever it is that makes them what they are!

Suppose, and here I anticipate a discussion to come,[32] into one of the gametes from which I developed a set of genes coded for repair enzymes had been inserted. These would stay inert but if and when I suffered from nuclear radiation these repair enzymes would correct the radiation damage as it occurred. If I then became exposed to radiation might I not say, and say correctly, thank goodness my parents arranged to have those genes inserted into me?

In the next chapter we will be considering cases of wrongful life, cases in which an individual may have been damaged by something done to her at the gametes stage. Here again it is natural to describe what has happened in just these terms: as something that happened to her at the gametes stage. Surely it is as natural as describing damage that happened later, as damage to her at the zygote stage?

What then is Williams's objection to the Gametic Principle? There seem to be two and they parallel the two senses of the term *unique* we distinguished earlier. On the one hand Williams is prepared to be pragmatic and revise the Zygotic Principle so that: 'the "earliest item" from which A uniquely developed is the zygote' except where there is twinning. In that case the 'history of A goes back to the splitting, and no further'. But we could say that this is simply false in a number of important senses. The history of A goes back to each of the gametes from which A derived and can, as we have seen, be crucially affected by what happens to those gametes.

True, the uniqueness of A is not fully determined by a single gamete, but neither is the uniqueness of A determined by all 46 chromosomes, which for many is the importance of conception, as the point at which the complete genetic structure of the individual is present in one place for the first time. All sorts of other things play a part, from the extrachromosomal mitochondria which are largely in the female gamete and which also carry DNA (but which may also originate from the male gamete and form part of the zygote having entered the egg literally 'on the tail' of the sperm), to the sort of nourishment A receives in the womb, to her toilet training, and so on. So why not say that the history of A goes back to each of the gametes from which A developed?

The only remaining reason is the possibility of the gametes resulting in more than one individual and therefore figuring in two different life stories. We could cope with this problem as Williams did in the case of the Zygotic Principle, by stipulating, 'Where there is twinning, the history of A goes back to the splitting, and no further'.

But why not further even in the case of twinning? Why this mystical infatuation with the individual? Why not accept the truth, that some individuals share part of their history with other individuals, that in some cases the same thing, *exactly the same thing*, happened to A *and to B* not simply simultaneously, but when A and B were one and the same individual?

Well, one objection to this, which doubtless is what Williams has in mind, is that it would then be true that A is identical to B and B is identical to C but A is not identical to C—but this cannot be true. It is this that impels Williams to insist that identity in the case of twinning can go back to twinning but no further.

But if this is right then an equally embarrassing problem arises. An individual who has no monozygotic twin can rightly say that she began at the gametes stage and trace her life history back to the gametes that formed her, but her friend who is a monozygotic twin cannot. Two individuals Jack and Jill are exact contemporaries and friends. One, Jill, not being a twin can say her life history began at the gametes stage but Jack, being a monozygotic twin, started somewhat later.

Personal identity is a relationship

If we consider the case of the monozygotic twinning of Tom and Dick we might get something like this: Tom and Dick started their life stories as two sets of gametes. These fused at conception to become a zygote, the zygote then split to form two separate zygotes which eventually developed into Tom and Dick. If this is what happened then it looks compellingly as though:

> The history of Tom and Dick is the history of their relationship to their DNA in the form in which it makes them the persons they are and will become.

This is the history of the formation, fusion, and subsequent fission of a cell or cells containing a particular DNA sequence and the relationship of this DNA to the human body or bodies, brain or brains, it generates.

Monozygotic twinning is a real and not uncommon event and it is like the science fictional cases of fusion and fission described by Derek Parfit in his celebrated account of personal identity.[33] For our present purposes we need not examine Parfit's arguments at any length except to record that Parfit believes that what matters is the holding of Relation R. Relation R is 'psychological connectedness and/or psychological continuity' no matter what causes this continuity.[34] That is to say, someone shares personal identity with another if they have an appropriate degree of psychological connectedness and/or psychological continuity with that 'other' individual. This allows for the possibility of successive selves and parallel selves. That is it allows for the person, Derek Parfit, to have many related incarnations, including, theoretically, parallel as well as subsequent incarnations. As Parfit put it:

> Though I do not survive My Division, the two resulting people are two of my future selves. And they are as close to me as I am to myself tomorrow. Similarly they can each refer to me as an equally close past self. (They can share a past self without being the same self as each other.)[35]

Parfit made psychological connectedness and/or continuity central because he was concentrating on personal identity that is the identity of persons properly so called. What these are we shall come to in a moment. Here I am interested in the identity of pre-persons. We need then both a concept of *personal identity* and one of *pre-personal identity*.

Pre-personal identity

Parfit in part developed his conception of personal identity to cope with theoretically possible cases in which adult persons might split or fuse—where for example a single brain might theoretically be divided and be transplanted into two different bodies with each half functioning as a full brain. He made the psychological continuity and connectedness of the consciousness involved central to the account that we would give of the identity of the various resulting individuals. However, the real possibility of monozygotic twinning and the reality of the fusion of sperm and egg show that we also need a conception of pre-personal identity. The most compelling is, I believe, the one we have arrived at here. Namely that:

> pre-personal identity is the relation between individuals and their DNA.[36]

The term 'pre-person' encompasses the zygote and the embryo and fetus right up to the point at which the person proper emerges, the person capable *inter alia* of psychological connectedness.

The history of the pre-person is the history of the related individuals which first combine in conception and then may again fuse in monozygotic twinning, and these individuals are related to the persons properly so called which will develop from the gametes and pre-persons.

When does a person's life story begin?

It seems clear to me that the life story of each person has two separate beginnings and must be traced to each of the gametes from which that individual developed, and that the history of each gamete is part of the history of any and every individual that develops from it.

There is of course also a moral argument for accepting the Gametic Principle. It is that it would be not only inconsistent but wrong to think that damaging the zygote or the embryo was wrong because it harmed an individual and the resulting person, and damaging the gametes was not wrong because it literally harmed no one! Of course neither is wrong if neither the zygote nor the gamete will be permitted to develop into a person, but both are wrong and equally so if it is known, or expected or even likely, that a person will develop from the damaged zygote or gamete.

Of course we could say that damaging the gametes is wrong not because it harms an existing individual but because, like damaging the ozone layer, it harms a future person or persons. While we *could* say this it seems more correct to say that in damaging the gamete we damage the person who develops from that gamete and we damage her at the gamete stage.

A person affecting morality cannot ignore the gametes because the gametes contain and transmit an important part of what the person who develops from those gametes will be like.

What's in a name?

A conception of identity which identifies identity as a relationship has a problem. This problem is what to say about the identity of the individuals related by the relationship. Describing the brain division case we have considered, the case in which two halves of a single brain are transplanted into two different bodies, Derek Parfit put the problem succinctly.

> In the case where I divide, though my relation to each of the resulting people cannot be called identity, it contains what fundamentally matters. When we deny identity here, we are not denying an important judgement. Since my relation to each of the resulting people is about as good as if it were identity, it carries most of the ordinary implications of identity.[37]

I think this is right, and because it is unrealistic to suppose we can effectively change normal usage to accommodate the distinctions involved in pre-personal identity I shall continue throughout this book to talk as if each of us has an identity with our gametes. However, nothing of importance turns on this. When I claim that the history of each of the gametes is an important part of the history of the person who develops from them, this is literally true and does not depend upon the person being identical with the gametes from which they developed.

However, when I say, as I do and will continue to do frequently, that people are or can be importantly influenced by what happened *to them* at the gametes stage, this seems to imply that one and the same individual is involved whereas strictly speaking we are talking of earlier and later selves, importantly related but not identical. Since I agree with Parfit that nothing of importance turns on this I shall continue to use the familiar locutions.

If all this is right, then Kripke's Zygotic Principle needs to be replaced by the Gametic Principle. Which brings us at last but briefly to the question of what is a person?

What sort of beings are persons?

I cannot attempt a complete account here of the point at which an individual becomes a person, but I can sketch the lines of such a defence. Most current accounts of the criteria for personhood follow John Locke in identifying self-consciousness coupled with fairly rudimentary intelligence as the most important features. My own account[38] uses these but argues that they are important because they permit the individual to value her own existence. The important feature of this account of what it takes to be a person, namely that *a person is a creature capable of valuing its own existence*, is that it also makes plausible an explanation of the nature of the wrong done to such a being when it is deprived of existence. Persons who want to live are wronged by being killed because they are thereby deprived of something they value. Non-persons or potential persons cannot be wronged in this way because death does not deprive them of anything they can value, though this does not exhaust the wrong that might be done by infanticide. Infanticide may still be a wrong against those who care for the intant—his parents or the Sisters of the Embryo for example —but it is not a wrong done to the infant. If they cannot wish to live, they cannot have that wish frustrated by being killed. So that creatures other than persons can be wronged in other ways, by being caused gratuitous suffering for example, but not by being painlessly killed. This explains the difference between abortion, infanticide, and murder and allows us to account for how we benefit persons by saving the lives of the human potential persons they once were, but at the same time shows why we do not wrong the potential person by ending that life, whether it be an unfertilised egg or a newborn infant.

On this account, the life-cycle of a given individual passes through a number of stages of different moral significance. The individual can be said to have come into existence when the egg is first differentiated or the sperm that will fertilise that egg is first formed. This individual will gradually move from being a potential, a pre-person into an actual person when she becomes capable of valuing her own existence. And if eventually she permanently loses this capacity, she will have ceased to be a person.

We can now return to the two women in Parfit's example and to the issue of the ethics of pre-natal screening.

The avoidance of needless suffering

Consider now a third woman. She is eighteen weeks pregnant and has an amniocentesis test. The results show that her fetus has spina bifida and she is offered a termination. She decides to terminate her pregnancy and try again for a healthy baby.

According to the arguments so far developed this woman acts in a way which is morally equivalent to the actions of the other two women. This is of course on the assumption that the abortion can be carried out without pain to the fetus. Like the second woman she decides not to actualize a potential person who will be injured, in order to have a healthy child at a later date. Both choose adversely to affect a potential person in being, for the sake of a future person. If we assume that all three women eventually have healthy children then the moral consequences of the policies adopted by all three women are the same.

If we want to say that the three women considered so far have all acted rightly and indeed that the courses of action chosen by each are morally equivalent must we also say that they would be wrong not to have acted as they did? Clearly the answer must be 'yes'.

If each of the three women has acted rightly, how do we characterize the actions of the three women? Each and all of them have chosen to try to bring into the world a healthy rather than a handicapped child when it appeared to them that they had a choice.[39] What then is the wrong of bringing a handicapped child into the world that these women and countless real counterparts choose to avoid doing so if they can, and regard themselves as acting for the best?

I want to suggest that the wrong they all try to avoid is the wrong of bringing needless suffering into the world. Each decided that when faced with the choice between having a healthy or a handicapped child they should choose to have a healthy child. Each could satisfy their desire to have a child without bringing into the world a child that would necessarily suffer. None wronged anyone, all benefited someone.

But when is suffering needless? Consider now the case of a fourth woman.

The fourth woman

The fourth woman is pregnant and is told that the child she is carrying will be born with a disability; moreover, that unfortunately any future children she has will also be born with the same disability. If she is to have children at all they will have disabilities and no pre-natal or post-natal treatments are available to ameliorate the disability.[40]

Are our intuitions the same? Do the good arguments that the mothers in Parfit's first two cases had for taking steps to avoid bringing a child with disability into the world still hold for the mother who can only have disabled children?

First we should notice that this woman cannot claim that an abortion would be in the best interests either of the child she is carrying or of any child that she will have in the future. For so long as the disability is not so great or so terrible that it would be better for this potential person or indeed anyone with such a disability that she had never been born, then it is in the interests of the child she is carrying to be born. Moreover, this woman can benefit no one (but herself and maybe her immediate family) by avoiding pregnancy.

So that if this fourth woman decides to go ahead and have her child even though it will be disabled, she is acting in the best interests of that child. Moreover, unlike the other three women, she would not be doing wrong in bringing avoidable suffering into the world. Any suffering that her child experiences due to its disability is unavoidable just in the sense that she can have no child that will not be subject to such a disability and whatever suffering goes along with it. This fourth mother wrongs no one in having a child with disability because it is in that child's interests to be born, she benefits that child by continuing her pregnancy and although the child will inevitably suffer, it will have a life worth living and such suffering as it will experience is unavoidable. Like the first three women she wrongs no one and benefits someone.

This case is puzzling, however. For the child is born with an injury. Knowingly bringing the child to birth with an injury that could have been avoided, this mother has injured her child. But, I want to suggest that this is a case where injuring someone does not constitute wronging them. The mother chose to injure her child but did not thereby wrong the child because the injury, *ex hypothesi*, would be worth having in exchange for existence. We could say the child would have been willing to accept the injury as the price for existence. This case is parallel to injurious side-effects to beneficial medical procedures. The injurious side-effects, if worth enduring for the sake of the benefits, are not thereby rendered non-injurious but they are not wrongs inflicted on the individual (unless of course inflicted against his will). And of course where a doctor must embark on a procedure which saves life at some cost in terms of injury to the patient, but cannot first obtain consent because, perhaps, the patient is unconscious, then if she goes ahead she will not thereby have wronged the patient. We will return to this point when considering the equally puzzling cases of 'wrongful life' in the next chapter.

We have one final case to consider, that of the woman who chooses to abort her pregnancy but not to have further children.

The morality of abortion

This case is easily dealt with. We do not need to consider why this woman wants an abortion. There is no risk of disability, and she will not 'replace' her aborted child with another at a later date. This mother wrongs no one and benefits no one. She wrongs no one because in ending the life of the fetus she deprives the fetus of nothing that it can value and she benefits no one since there is no one she brings into existence.[41]

Interim conclusion

We can then come to an interim conclusion about the morality of selective termination and more generally about the morality of bringing individuals into the world. I want to suggest that the rationale that unites and explains our intuitions about the various cases we have considered is that it is wrong to bring avoidable suffering into the world. One consequence of this conclusion is of course that the decision to go ahead and have a child requires as much and as careful justification as the decision to terminate a pregnancy and that it can also be wrong not to terminate a pregnancy.

We have recognised the powerful desire and the strong interest that, people generally have in having children. Just as this desire should be exercised responsibly we should also be careful not to frustrate it without good reasons. If children are wanted, it is better to have healthy children than to have disabled children where these are alternatives, and it is better to have children with disabilities than to have no children at all.

In suggesting that it is wrong to bring avoidable suffering into the world and in indicating that suffering is avoidable where an individual who is or will be disabled can be replaced with an individual who is not disabled I have assumed that replaceability is unproblematic. That is to say, I have assumed that if a woman delays conception or terminates one pregnancy in order to initiate another pregnancy later, this course of action will be successful. Now clearly many things can and do go wrong with pregnancy. Miscarriage is common and a host of other problems are more or less probable in any pregnancy. If, for good reason, it appears to a particular woman, perhaps because she is nearing the end of her childbearing years, that she is unlikely to be able for example to replace an established pregnancy with any reasonable probability of success, then in such a case if the child she is carrying will be disabled it is surely reasonable for that woman to be profoundly sceptical about the possibility of replacing her child with a subsequent more successful pregnancy and so she is surely entitled to regard that disability as unavoidable.

It is important to be clear that where we do choose to bring avoidable suffering or injury into the world this is wrong. But unless the injury or suffering to the individual thus created is so great as to make life intolerable then this individual is not thereby wronged. The wrong is that of an individual deliberately choosing to increase the suffering in the world when she could have avoided so doing.

Existential value

Before continuing I must make one point very clear. None of this means that healthy, or unimpaired, or non-handicapped, or non-disabled children are better in any existential sense than those with disability or handicap.

If I say, as indeed I would, that I would prefer not to lose, say, a hand, that it would be better for me if I did not lose one of my hands, that I would be better off with both hands and so on, I am not committing myself to the view that if I did in fact lose a hand that I would therefore and automatically become less morally important, less valuable in what I am calling the existential sense, more dispensable or disposable than you. I have a rational preference not to lose any of my limbs, I have a rational preference to remain non-disabled, and I have that preference for any children I may have. But to have a rational preference not to be disabled is not the same as having a rational preference for the non-disabled as persons.

II. Screening Would-Be Parents

We have looked at the circumstances in which children or potential children are screened when questions arise as to whether it would be ethical to bring them into the world or to allow them to continue in the world. We must now look briefly at a parallel case in which it is the parents who are screened for their adequacy as parents. But unlike pre-natal screening the parents are screened not, so to speak, for their fitness to exist or continue to exist as persons but simply as parents. The concern here, as in pre-natal screening, is with the quality of life of the children who are or may be in their care.

There are a number of different sorts of cases in which such screening takes place. There are the sorts of cases that have come before the courts in which very often it is mentally handicapped girls who are subjected to compulsory sterilization or even compulsory abortion. These cases raise difficult issues, some of which have to do with autonomy and consent, some of which have to do with the rights of the handicapped, and some of which turn on the question of what constitutes reasonable circumstances into which a child might be permitted to be born. All of these are too difficult to discuss in any detail in the space and time remaining, and many of these issues have been addressed at length elsewhere.[42]

There are of course also the cases of routine screening of adoptive and foster parents which are far from unproblematic and raise fundamental issues of principle which are seldom discussed. And which again will not be discussed in their peculiar detail now.

Equally, those whom one might term 'established parents', those who have had children, are often screened by social services and other individuals and agencies with the possibility of losing custody of their own children. And recently there has been at least one case where such screening by social services has been carried out on a woman of eight months' gestation who was at that stage judged unfit as a potential parent and warned that her child would be taken into custody when it was born.

All of these cases raise fundamental issues about the right to found a family and the criteria for adequate parenting. However, one increasingly common form of pre-natal screening of parents has been little discussed and it is to this that I would like to devote rather more attention in the remainder of this chapter before trying to draw some general and I am afraid very swift conclusions about the legitimacy of screening parents and selective termination of the right to found a family.

It is common for those administering *in vitro* fertilization (IVF) programmes to screen parents as to their suitability. Not only as to their suitability to benefit medically from the procedures, but as to their suitability as parents *per se*, and as to their moral claims on a scarce resource. It is instructive to look at the criteria employed by one major centre for IVF. This centre was the first National Health Service clinic providing IVF and was established at St Mary's Hospital, Manchester. The following criteria are those employed at this clinic and there are points in common with those in use at other centres around the country. I should make clear that these are the criteria for being admitted onto the waiting list for IVF and it is of course not certain that those admitted to a waiting list will eventually be treated.

1. Women must be less than 36, men less than 46.
2. Couples must have lived together for three years.
3. No children living with the couple.
4. No major physical or psychiatric illness.
5. Evidence of regular ovulation.
6. Couples not accepted if a male factor is the sole cause of infertility.
7. Ovaries accessible to laporoscopy or ultrasound.
8. Live in North West Health Authority Region.
9. Female close to ideal body weight for height.
10. Must not have had more than two complete courses of IVF or 'Gift' treatment elsewhere.
11. Must fulfil adoption criteria.

I should make clear that only heterosexual couples are considered but of course by no means all of these. Excluded are many heterosexual and even married couples, couples who may have reached the far edge of their fertility, single parents, and gay couples, and of course couples who have children living with them, even if those children are not their own genetically.

Each of these criteria is worthy of detailed examination and each raises many interesting issues of principle. However, it is clear both from the criteria themselves and from the gloss that has been offered on them,[43] that they are designed to achieve three separate aims. First, there is the aim of selecting among superabundant candidates for a scarce resource. Second is the aim of gaining maximum benefit in terms of successful outcomes (children) from the use of that resource; and finally there is a concern with the quality of parenting that would result.

While it may seem on the face of it that each of these are legitimate and even worthy aims, let alone essential ones in the case of allocating a scarce resource, I believe that they are either illegitimate as aims or as means of achieving legitimate aims. Having made these extravagant claims I am going to attempt to substantiate only one of them.[44]

Should we screen parents?

The eleven criteria used for selection of candidates for IVF reveal a number of assumptions about adequate parenting. These can be separately identified as follows:

1. Adequate parenting requires two parents, one of each gender.
2. Adequate parenting requires parents to be under 65 years of age while the children are still at home.
3. Adequate parenting requires that neither parent have major physical or psychiatric illness.
4. Adequate parenting requires stable relationships between the parents, evidenced by three years' cohabitation.

I know of no evidence for the truth of any of these criteria. Indeed I know of no reliable evidence for any criteria of adequate parenting that can be applied to potential parents rather than to actual parents who have proved their inadequacy in objective ways. Indeed some of these criteria are scarcely plausible.

But suppose for a moment that there was sufficient evidence in favour of these criteria to make their use defensible. If this were the case then we should certainly not only apply them when candidates for a scarce resource come forward. We should apply them to any and all candidates for medical assistance with reproduction. General practitioner advice is probably the commonest form of medical assistance with procreation and it should certainly be withheld from anyone not meeting defensible criteria. As should the prescription of drugs such as clomiphene and other methods of assistance with ovulation or fertility. The use of HCG and other drugs to support established pregnancies should be withheld from unsuitable candidates for parenting and so on. Of course in the case of general practitioner advice, the patients can just go ahead and attempt to have children anyway, but in so far as the advice was calculated to facilitate successful pregnancy or make such an outcome more likely it should in consistency be withheld. Of course much advice is not on how to establish a pregnancy but how to minimize dangers to the child and such advice should of course continue to be available.

It is irrelevant that these resources are cheap and plentiful when compared with IVF for if the grounds of selecting candidates are their adequacy as potential parents, this does not improve in proportion to the cheapness of the treatment. But, more significant still, if we had any confidence at all in this or indeed in any other prospective criteria of adequate parenting, we would be wrong to apply them only in these marginal cases. If we were at all serious about preventing inadequate parenting and permitting only sound potential parents to procreate, we would not, as we do, let any Thomasina, Dorothea, or Harriet go out and have children without so much as a 'by your leave'. Remember that screening candidates for IVF let alone adoptive or foster parents is a tiny and numerically insignificant proportion of the population of parents.

The alternatives seem to be these. Either we believe that we have adequate and defensible criteria for adequate and inadequate parenting and it matters that people demonstrate their suitability for parenthood in advance, or it does not. If it does, then we should license parents, all parents. If it does not, then we should not do what amounts to victimization and apply indefensible criteria to the few unfortunates that need assistance with procreation.

The reason we do not in fact attempt to license all parents is not because we could not, but because there are good reasons not to. Some of these good reasons have to do with the inadequacy of speculative criteria about good parenting. But more importantly they have to do with the importance and value that most people attach to the freedom to have children coupled with our reluctance to place the comprehensive powers that licensing would involve in the hands of anyone at all, whether that person be some central authority or an individual doctor or social worker.

But these powers would be unnecessary for we already have the only reliable method of protecting children from inadequate parenting. We remove children from the custody of parents who have palpably ill treated or placed in danger their children, and disqualify potential parents who have proved their unfitness by a history of damaging or mistreating their children. While this is by no means easy to achieve it is the only defensible method.

In short there is only one reliable criterion for inadequate parenting, it is the palpable demonstration of that inadequacy in terms of cruelty, neglect, or abuse of children.[45]

Conclusion

Just as in the case of screening potential children for their adequacy as children we found that while happy and healthy children were better than unhappy or handicapped ones, handicapped children were better than no children at all, so in the case of screening parents. After all, the justification for screening parents is the same as the justification for screening children. It is that we ought not to bring avoidable suffering into the world.

Now parents can very often choose to replace the child who would have been born handicapped with one who will not, whereas children cannot replace defective parents so easily—not at least when those children are not yet in existence.

The important differences between pre-natal screening of children and pre-natal screening of parents seem to be these:

1. There are no reliable predictive criteria for inadequate parenting.
2. While there is evidence that children do suffer disadvantage from sub-optimal parenting when that is defined in socio-economic terms, the therapy of choice would be to work on the socio-economic conditions rather than depopulate the world.
3. Even if there were reliable predictive criteria of good parenting, the moral consequences of denying parents equal opportunities to reproduce are worse than the consequences of allowing some less than optimal parents to reproduce.

If parents want children of their own they should have every assistance with childbearing which is consistent with the like assistance for all other citizens. This means, where resources are scarce, distributing those resources in ways which do not discriminate against parents on the grounds of dubious allegations as to their adequacy as parents.

To fail to carry out pre-natal screening and to decide to decline selective termination where the circumstances make it inevitable that avoidable suffering will thereby be brought into being is to act wrongly precisely because it causes avoidable suffering. Similarly, if we screen parents for their adequacy as parents in advance of their being or becoming parents, then since there is no evidence that this prevents suffering to children and there is plenty of evidence that it causes suffering to the potential parents this is also wrong and for the same reasons, namely that it causes avoidable suffering.

4

The Wrong of Wrongful Life

The idea that it might be a moral crime to have a baby, that it might be wrong to bring a new human individual into the world is to many people simply bizarre. Having a baby is a wonderful thing to do, it is usually regarded as the unproblematic choice from the moral if not from the 'social' or medical point of view. It is only having an abortion and perhaps also refraining from having children that is regarded as requiring justification. However, the idea that bringing a child into existence might be actually wrongful as opposed to simply inconvenient, embarrassing, dangerous, or distressing for the parents is not new. As we noted earlier,[1] John Stuart Mill regarded bringing children into being without the prospect of adequate physical and psychological support as nothing short of a moral crime.

The crime of having children

Of course, the crimes to which Mill alludes are relatively easily remedied, as Mill is himself aware. For he glosses his indictment thus: 'if the parent does not fulfil this obligation, the State ought to see that it is fulfilled, at the charge, as far as possible, of the parent'.[2] It is precisely these ideas that are at the heart of the highly controversial and philosophically extremely interesting so-called 'wrongful life' law suits that have recently proliferated in the United States and are also appearing in the United Kingdom.

Wrongful life

The idea of 'wrongful life' is simply that an infant has been harmed and/or wronged by being brought to birth in a less than satisfactory condition or in adverse circumstances. Most of the cases to date involve very serious harmful conditions, although, interestingly, the term 'wrongful life' was first used in a case in which a healthy infant claimed that he had been injured by being born to less than optimal parents, in this case, that his father had allowed him to be born illegitimate.[3]

Initially such actions did not succeed. For example, *Gleitman* v. *Cosgrove*[4] involved a child who had been born a deaf mute and almost blind because his mother had contracted German measles during pregnancy, but the Supreme Court of New Jersey would not accept the plaintiff's claim for damages against doctors who allegedly told the mother that German measles afforded no risk to her child.

However, the courts of California, Washington, and New Jersey have all now recognised the right of an infant with birth defects to collect damages in a wrongful life suit. The California case, *Curlender* v. *Bio-Science Laboratories*,[5] for example involved a child born suffering from Tay-Sachs disease. The parents were awarded damages for having negligently been told they were not carriers of the disease.

The claim in such actions is brought by or on behalf of the child and, as Bonnie Steinbock has explained:

> The claim in a wrongful-life suit is not that the negligence of the physician was the cause of the impairment. It is, rather, that the physician, by failing to inform the parents adequately, is responsible for the birth of an impaired child who otherwise would not have been born and therefore would not have experienced the suffering caused by the impairment.[6]

There is a distinction which we should note between 'wrongful birth' and 'wrongful life' cases. In the former, it is usually the parents who bring an action against a physician for negligence, the result of which is that they, the parents, have been deprived of an option to abort. 'Wrongful life' on the other hand refers to actions brought by or on behalf of infants for damages arising from simply having been born.[7]

We should also note that 'wrongful life' cases differ from those in which it is alleged that a mother may have damaged her child *in utero* by, for example, taking drugs or indeed by excessive alchohol or cigarette smoking during pregnancy. Such harms to resulting individuals as may flow from this conduct, while significant and interesting, are beyond the scope of our present concerns although they do of course concern analogous harms and wrongs.

What is at stake?

The idea that one might be harmed or wronged by being brought into existence in a less than satisfactory state is very important indeed, for, as we have seen, it challenges many of our moral presumptions about having children. Moreover, if the alleged wrong can give rise to legal actions for compensation, and perhaps also to criminal liability, then a number of further problems arise. All of these further problems are the subject of this chapter.

First, we will have to consider whether in the light of such possibilities we are prepared to accept that the criminal or civil law is a fruitful or indeed an ethical mechanism to use, either as a way of attempting to reduce the number of children born with disability or as a way of compensating those who are so born. Moreover, so long as litigation is being used for these purposes, the identification of the precise nature of the alleged harm and wrong is crucial.

An initial problem here is of course to have some sense of just how unsatisfactory the condition of a child must be before it can claim to have been wronged or harmed by being brought to birth. At the moment it is infants with birth defects or impairments that bring such suits or have them brought on their behalf, although as we have seen, even individuals who have supposed themselves to have been disadvantaged by illegitimacy have attempted to gain compensation for their supposed unsatisfactory state.

A further and new problem is on the horizon. With the rise of bioengineering we have the possibility of genetic manipulation being used to confer substantial advantages on engineered individuals or to remove substantial impairments, like susceptibility to disease, or, perhaps, the particular gene responsible for a genetic disorder. It may well be that children who today would be regarded as normal healthy children will in the future make claims that they have been wronged or harmed by being denied genetic enhancement or genetically engineered removal of impairment. To know whether this might ever be the case we must know precisely how to characterise the alleged wrong of wrongful life. For the crucial issue will be whether such individuals can validly claim that their condition may plausibly be regarded as constituting a harm to them or that they have been wronged by being brought to birth in such a state.

For these reasons we must turn first to the problem of characterising the supposed wrongs at stake.

What's wrong with life?

Some United States judges have regarded claims for wrongful life as too logically puzzling to be sustained. As the judge in *Gleitman* v. *Cosgrove* put it:

We must remember that the choice is not between being born with health or being born without it; it is not claimed the defendants failed to do something to prevent or reduce the ravages of rubella. Rather the choice is between a worldly existence or none at all. To recognise a right not to be born is to enter an arena in which no one could find his way.[8]

In the principal English case on wrongful life, *McKay* v. *Essex Area Health Authority*,[9] the Court of Appeal took the same line saying that 'the difference between existence and non-existence was incapable of measurement by a court'.[10] Ackner LJ, in his judgement in that case, also said that he could not accept that

the common law duty of care to a person can involve, without specific legislation to achieve this end, the legal obligation to that person, whether or not in utero, to terminate his existence. Such a proposition runs wholly contrary to the concept of the sanctity of human life.[11]

A leading English legal writer, Margaret Brazier, has noted that the 'Court of Appeal in McKay, said that the difference between existence and non-existence was incapable of measurement by a court. The difference between the cost of bringing up a healthy child and the cost of bringing up a disabled child can be measured with some degree of accuracy'.[12] The same conclusion has been reached by J. K. Mason and R. A. McCall Smith who conclude that 'The comparison to be made is not that between non-existence and a deprived life but that between a defective life and one of a normal child'.[13]

While this may well be the right comparison to make in deciding what the child needs by way of compensation for being born disabled, the problem of knowing whether this disability may rightly be attributed to the actions, whether wrongful or not, of others remains. As does the equally pressing problem of determining whether the child may justly be said to have been injured or wronged by being brought to birth.

Steinbock's view

In her illuminating account of the whole issue of wrongful life, Bonnie Steinbock rightly rejects the spineless approach to the problem exhibited in *Gleitman* v. *Cosgrove* and then reviews and rejects a number of the ways in which the courts of the United States have attempted to rationalize their solutions. She finally recommends what she calls 'the correct interpretation' of the way to characterize the infant's injury. The problem is to characterise just what injury an infant who has a restricted life, though still one worth living, has suffered by being born. Steinbock follows her compatriot Joel Feinberg in suggesting:

> Talk of a 'right not to be born' is a compendious way of referring to the plausible moral requirement that no child be brought into the world unless certain very minimal conditions of wellbeing are assured. When a child is brought into existence even though those requirements have not been observed, he has been wronged thereby.[14]

This is of course resonant of the passage from John Stuart Mill[15] quoted above and relies on the idea, implicit in Mill and outlined above, that it can be a moral crime to do things to the pre-person (the human individual at any stage prior to the onset of fully fledged personhood whenever that is deemed to occur) or indeed to the possible person, including causing that individual to exist.

Steinbock interprets Feinberg's central idea as expressing the judgement that it is a wrong to the child to be born with such serious handicaps that many very basic interests are doomed in advance. Steinbock insists that 'While this is something less than a right to be born a whole functional human being, it is not dependent on the implausible view that a life with serious impairments is always worse than no life at all'.[16]

Steinbock concludes that when very many of a child's basic interests are doomed in advance then wrongful life suits should succeed. Steinbock is attracted to this solution because she regards it as avoiding the embarrassment of having to compare existence with non-existence. It also avoids the severity of having to say that an individual's life must be so terrible that continued existence is worse than death before it can be justifiable to conclude that such an individual might plausibly be said to have been wronged by being brought into existence. The severity of such a view lies in the brutal reality that under current government policies in most of the world, and certainly in the United States and Great Britain, the disabled individual can only get compensation through litigation.

There seem to me to be two problems with Steinbock's approach. One is that it is something close to the view that Steinbock calls implausible that is at the centre of Feinberg's position and on which she partly relies. This is the view that life with serious impairments must be worse than no life at all before such a life can be said to be wrongful.[17] The other is that both Steinbock's position and Feinberg's take a somewhat perverse view of the distinction between harming and wronging.

For Feinberg, the child cannot claim to have been harmed unless it has been made 'worse off', and it cannot be worse off for having been brought into existence unless existence is, for that child, so terrible as to be not worth having. If this is right almost all wrongful life suits would fail, for most disabled children have lives that, while restricted and blighted in many ways, are still worth living. It is this fact, which means that such children have little hope of meaningful financial support and compensation, that leads Steinbock to champion a less restrictive view. But can such a view be sustained? To see whether it can we need to examine Feinberg's arguments at greater length.

Feinberg's view

Joel Feinberg, in a major philosophical discussion of the idea of a wrongful life[18] introduces a complex set of distinctions between what might count as 'harming', what might count as 'wronging', what might count as 'doing wrong', and in what circumstances these might legitimately interest the criminal law. One very important consequence of the way that Feinberg draws these distinctions, as we shall see, is to take the idea of 'wrongful life' out of the domain of the criminal law. It is likely that the way he draws the distinctions, if accepted, would also take 'wrongful life' beyond the reach of civil liability, except in the rarest of circumstances, as the law in most societies presently stands. In suggesting (1) that no harm has been done to the victim of 'wrongful life', and (2) that no wrong has been done unless the consequences for the 'injured' (?) party are 'so severe as to render his life not worth living', Feinberg leaves the injured party with nothing legally to complain of.

Feinberg clearly sets out his conclusions as follows:

1. In wrongfully conceiving the child despite the known risk of genetic deformity (say), A and/or C (the biological parents) do not harm B (the resultant infant), even if B comes into existence in a state that makes a 'life worth living' impossible. B has been born in a condition extremely harmful to it, but strictly speaking, that is not a harmed condition, not the effect of a prior act of harming. To be harmed is to be put in a worse condition than

one would otherwise be in (to be made 'worse off'), but if the negligent act had not occurred, B would not have existed at all. The creation of an initial condition is not the worsening of a prior condition; therefore it is not an act of harming, no matter how harmful it is.

2. Nevertheless, the wrongful act of A and/or C can wrong B even though, strictly speaking, it does not harm B, provided its consequences for B are so severe as to render his life not worth living. In that case B comes into existence with his most basic 'birth rights' already violated, and he has a genuine moral grievance against his parent(s). To the parent's defence that the only alternative to so harmful an existence was the nonexistence that would have followed their abstinence, the infant can make the rejoinder that nonexistence was the preferred alternative, even though it would not have been a 'better off' condition of B. Any rational being, he might add, would prefer not to exist than to exist in such a state.[19]

Feinberg insists that the child has only been wronged where non-existence is preferable to her present condition and that she has not been harmed at all, for the simple, and to Feinberg sufficient, reason that she has not been made 'worse off'. Moreover, a less severely impaired child who would not prefer never to have been born has not on Feinberg's view been either wronged or harmed. And this will be true, even though its disabilities might be very substantial, so substantial as to put the individual in a 'severely harmed' state. A mother who deliberately has a child knowing this will be the case is still, however, a wrongdoer although, and paradoxically, she has wronged nobody. Her wrong would, according to Feinberg, be that of 'wantonly introducing a certain evil into the world, not . . . inflicting harm on a person'.[20]

Feinberg concludes that if

this result strikes the liberal reader as counterintuitive, his best recourse is to modify the harm principle so that it accepts as a reason for criminal prohibitions not only the need to prevent people from wrongfully harming other persons, but also the need to prevent people from wrongfully bringing other persons into existence in an initially harmful (handicapped) condition.

and Feinberg adds: 'After careful consideration however the need for such a revision is not apparent to me'.[21]

There are a number of problems with the way in which Feinberg elects to analyse the idea of harm, to which we will come in a moment. Some of these stem perhaps from his preoccupation with the connection between harm and criminal liability. However, it is not clear that we would always wish to exempt 'wrongful life' from the criminal law.

Consider the case of selective termination. Where as a result of induced superovulation or of the implantation of a number of fertilized ova, a woman becomes pregnant with perhaps five or more embryos, selective termination is sometimes considered. Surplus embryos are deliberately killed leaving perhaps two or at most three survivors. This is because it is believed, on good evidence, that if all are allowed to go to term it is more likely that none will survive than would be the case if only two or three embryos are left in place. Suppose now that we can screen the embryos for genetic defects and select amongst them as to which to terminate. A mother or a doctor who deliberately chose to terminate the healthy embryos and allow the disabled ones to come to birth would it seems to me not only have harmed those disabled individuals who were born, but would have done a wrong that might appropriately interest the criminal law. Feinberg would, however, have to think not, because he wishes to limit the interest of the criminal law by employing the harm principle and on Feinberg's version of that principle no harm would be done to the resulting handicapped individual by either the doctor or the mother in this case. Indeed Feinberg would not only have to say that no one was harmed by terminating the healthy rather than the defective embryos but he would have to see nothing criminal in it either.

It is true that Feinberg would still regard what the mother and the doctor in such a case have done as profoundly wrong. He and I would agree that among the wrongs done is the wrong of 'wantonly introducing a certain evil into the world'. However, Feinberg thinks this is the only wrong done in such a case. It is the double insistence that neither the mother nor the doctor have done harm nor done anything which might legitimately be termed 'criminal' which seems perverse here.

The decision to 'criminalize' conduct is surely principally a question of the utility of so doing. The issue is most sensibly decided by weighing up the social, political, and moral consequences of using the apparatus of the criminal law and of imposing the stigma and social consequences of criminality on offenders. We should not predetermine this issue by deciding in advance that if conduct is not harmful it is not criminal.

Of course the notion of criminality complicates the issue unnecessarily. We may well want to stop short of criminalizing all sorts of wrongful acts including wrongful harmings. And Feinberg might, not unreasonably, simply accept the consequence of having to regard selective termination of healthy embryos in favour of disabled embryos as neither harmful nor criminal for the sake of the utility of his approach as a whole. However, there does seem to be something wrong with the Feinberg approach to harming which is highlighted by the case of selective termination.

If one is interested in 'the moral limits of the criminal law' (the subtitle of Feinberg's book) it is not unproblematic that a doctor and a mother who deliberately planned to bring into being handicapped rather than healthy children in the circumstances described, have done nothing that might at least prima facie reasonably interest the criminal law. Whether as a matter of public policy the criminal law is the best way of regulating such matters is of course a further and separate question.

Harming, wronging and criminality

It is important to be clear just what is at issue here. Feinberg wants to distinguish the ideas of 'harming' and 'wronging' and wants to confine the interest of the criminal law to the domain of harming alone, leaving wronging to the domain of the moral. For Feinberg the criminal law can be invoked to prevent or punish the doing of harm but not to prevent or punish the doing of wrong, unless that wrong also constitutes a harming.

Now Feinberg could claim that there is a sense in which any wrongdoing harms society—makes society worse than it was before—and so all wrongdoings can interest the criminal law. But Feinberg would surely never make such a claim for to do so would of course demolish his carefully drawn distinction.

So that if Feinberg's distinction is to prove useful it must plausibly delineate an area of harming which is significantly a special type of wrongdoing such that the criminal law might reasonably interest itself in wrongful harming but not in non-harmful wrongdoing. However, it seems obvious that the criminal law cannot plausibly be excluded where people have not been made 'worse off' in Feinberg's sense. There may be compelling social or political reasons why the criminal law might, in some circumstances, be interested in non-harmful wrongdoings. But more importantly perhaps, Feinberg's definition of harming is strongly counter-intuitive and must of needs give way to an account which is both more plausible and does the work required with more economy.

What is it to harm someone?

Feinberg insists that to be harmed is to be put in a worse condition than one would otherwise be in: that it is to be made 'worse off' and, in particular in these sorts of cases, worse off 'in the special narrow sense that requires both set-back interests and violated rights'.[22] It seems, on the other hand, more compelling and more cogent to say that to be harmed is to be put in a condition that is harmful. A condition that is harmful, Feinberg and I would agree, is one in which the individual is disabled or suffering in some way or in which his interests or rights are frustrated. The disability or suffering may be slight, just as harms may be trivial. And of course what constitutes a disability may have to be defined relative to a particular population. In a population which has been genetically engineered to be resistant to all major infections including AIDS, hepatitis, and heart disease, someone who has not been thus protected would be severely disabled.

I would want to claim that a harmed condition obtains wherever someone is in a disabling or hurtful condition, even though that condition is only marginally disabling and even though it is not possible for that particular individual to avoid the condition in question. For example I have a rational preference not to lose the little finger of my left hand, and I have such a preference that my children should be born with all their fingers. I have this preference because the condition, although relatively minor, is a disabling and/or a hurtful one. If I lose that finger or my child is born without it, I and she are to that extent in a harmed condition. If my daughter had no option but to be born without a little finger, if she suffered from a genetic defect that involved having only four digits on her left hand, then for her it was life thus harmed or no life at all. It was not possible for her to have all her fingers. But to be born thus, albeit slightly, disabled is to be born in a harmed condition and one that she could have a rational preference to be without.

I say 'could have' rather than simply 'has' because the individual may, for all sorts of reasons, not have such a preference. In the particular circumstances of the case, such a condition may be advantageous to the individual. In her nineties and looking back on a life drawing to its close, my daughter might be able to conclude that overall she has benefited from the lack of a finger. But it is still something disabling or hurtful, it still counts as being in a harmed condition.

So we can agree for present purposes, that a harmed condition is one in which the individual is disabled or suffering in some way. This enables us to come to an interim conclusion.

What it is to harm someone.

> Where B is in a condition that is harmed and A and/or C is responsible for B's being in that condition then A and/or C have harmed B.

In the case of wrongful birth, A and/or C have not only caused B to be in a condition which is harmful but are also morally responsible for B's being in that condition, therefore A and/or C have harmed B. A thus harms B whenever A puts B in a harmful condition. Where A is morally responsible for putting B in such a condition, then A is morally responsible for B's condition.

It might appear that the difference between Feinberg and myself here is both slight and, worse, merely semantic. Certainly we are not far apart but the small distance is significant and not 'merely' semantic though of course it is also, and among other things, semantic.

My point is simply this. Where someone has caused another to be in a harmed condition and is moreover fully morally responsible for having caused such harm, it is natural and logical to say that they have harmed that other person. To deny this requires a special justification. Feinberg's justification must be that using words his way confers some theoretical advantage either in clarity, or by enabling subtle and important distinctions to be made, or in some other way. In particular Feinberg wants to make plausible the exclusion of the criminal law to cases where individuals have wronged one another but caused no harm. My concern has been to show that neither the exclusion of the criminal law nor the strain of denying that harm properly so called has been caused, are either plausible or justified.

There is another point as well. It is that our normal sense of what it might be to wrong someone as opposed simply to harming them dictates that we draw the distinction in the diametrically opposed way to that preferred by Feinberg. And if we do so, we can the better see the appropriate place for the law in these questions.

Wrongful lives but not wronged individuals.

What then is the wrong of wrongful life? It can be wrong to create an individual in a harmed condition even where the individual is benefited thereby. The wrong will be the wrong of bringing avoidable suffering into the world, of choosing deliberately to increase unnecessarily the amount of harm or suffering in the world or of choosing a world with more suffering rather than one with less.[23]

Whether or not mothers and/or fathers should be blamed, and to what extent, for committing such a wrong is of course more problematic. And the further question of the extent to which they should be the subjects of litigation whether civil or criminal is even more problematic. I am inclined to think that where parents can avoid doing such wrong by declining to have the disabled child and yet fulfil their desire to have a child by trying again for a healthy child then if they deliberately produce children with more than slight disability they are blameworthy. If, however, the particular parents must have disabled children if they are to have children at all, then they will be blameworthy only if the children would be wronged by existence, that is, if they would find life worth not living. I offer these conclusions here simply to indicate some of the ramifications of the present discussion. My reasons for coming to these conclusions I have set out in detail elsewhere.[24]

For present purposes the important point is that the decision to have a child is not a wrong against the individual thereby created, if that individual has benefited overall by the decision or by the negligence. Now Feinberg would clearly accept the reasoning of this last sentence and this is, I think, why he wants to insist that no wrong is done unless it is plausible to regard the adversely affected individual as losing by the transaction. And this only happens in the case of creating an individual, where they would wish not to have been created at such a cost to themselves.

But this seems to turn the notions of harming and wronging upside down.

Harming and wronging

Adverse side-effects, of treatment for example, are adverse effects even if in the circumstances they are worth enduring for the net benefits. Those treating patients who cannot give consent must be mindful that harmful treatments, treatments with adverse side effects, should only be used where there is no equally effective treatment available with less serious side effects and where the treatment advantages the patient overall. But the need to be mindful of these things is explained by the fact that the treatment does cause harm. Indeed, Feinberg's own explanation of his conclusions seems to make the same point.

Feinberg at one stage of the argument wants to define the harms done to children allowed to come to birth with defects, in terms of that child's being deprived of 'birth-rights'. He explains that

> if before the child has been born, we know that the conditions for the fulfillment of his most basic interests have already been destroyed, and we permit him nevertheless to be born, we become a party to the violation of his rights.

Feinberg insists that the interests thus doomed must be very basic ones and concludes:

> to be dealt severe mental retardation, congenital syphyllis, blindness, deafness, advanced heroin addiction, permanent paralysis or incontinence, guaranteed malnutrition and economic deprivation so far below a reasonable minimum as to be inescapably degrading and sordid, is not merely to have 'bad luck'. It is to be dealt a card from a stacked deck in a transaction that is not a 'game' so much as a swindle.[25]

This is an astonishing passage. Since no one person could be expected to be born with all these conditions they must be intended as alternatives. Should congenitally deaf people who have happy lives decline to procreate and produce probably equally happy but deaf children, because to do so would be a swindle? Should the very poor not procreate at all? Of course such people choose to create harmed individuals but where the lives of the children, despite disability, may be expected to be well worth living it is hardly a swindle.[26] Even if we take a more charitable view of what is meant here and assume Feinberg intends some combination of these conditions we get a scarcely more plausible scenario.

But the passage is astonishing for another reason which bears vitally on the issue of wrongful life. If deliberately to choose to produce a child with any of these conditions and perhaps others like them or even a combination of them is to doom one of that child's basic interests to defeat, then that choice has surely harmed the individual. Indeed, Feinberg deliberately chooses the vocabulary of action and consequence, he talks of 'dealing from a stacked deck'. Feinberg of course sees this for he says:

> A child born with such handicaps is in a condition that we should not hesitate to call 'harmed' if it were not for the fact that it is not, like standard harms, a worsening of some prior condition . . . there is good reason to claim that he has been wronged to be brought into existence in such a state. (The state is a harmful one, even if it is not, strictly speaking, a 'harmed' one.)[27]

It is surely both natural and correct to say that the person who caused a harmed condition to obtain has caused harm, and to cause harm is, of course, to harm.

Beneficial self-harming

There is nothing illogical or suspect about saying also that it can be in my interests to be harmed. Harming may constitute a net benefit. Although again, Feinberg seems to balk at this. When in the First World War soldiers deliberately shot themselves in the foot, or injured themselves in some other way so as to get what was called a 'Blighty Wound', one that would get them sent home to 'Blighty', and out of the fighting, they were guilty of an act of deliberate self-harming. Indeed were it not an act of self-harming, which may have disabled or handicapped the individual to some extent, it would not have secured the desired effect.

There is of course a nice ambiguity here between whether or not the self-infliction of a Blighty wound renders the individual worse off. Clearly he is worse off than he was immediately before the wounding in the sense that formerly he was pain free and unwounded and latterly, presumably, he is in pain and wounded. But since his prospects of premature death have receded he is in another sense better off, better off overall. Insistence on tying harm to the idea of being made 'worse off' deprives us of the ability to characterise what is going on here as a self-interested act of self-harming. It is surely clearer and more consistent with what we wish to say in such cases to describe the acts of these soldiers as acts of self-harming but by which they did not wrong themselves.

Again it is clear that the opposite way of drawing the distinction to that preferred by Feinberg is both the clearer and more useful. Rational suicide, suicide in circumstances in which it is clearly in the subject's own interests to die, or voluntary euthanasia in the same circumstances will be the limiting case here.[28]

Now all this is by way of answer to Feinberg's own prior question: 'how severe must the harm be to be a "worse off condition" than nonexistence?'[29] But if this is the crucial question, there is an important difference between asking it of individuals who can have a view about the desirability of their own existence and asking it of others. Concerning children or adults in being who can have a view about the worthwhile nature of existence the question can surely only be answered subjectively. A condition is worse than non-existence if and only if the subject would rather not exist than exist in such a condition. If this is right all the talk of basic interests being doomed and so on is redundant. For Feinberg believes that it is logically impossible for an individual to have been wronged by being brought to birth unless it is true that 'nonexistence was the preferred alternative'[30]—preferred not preferable.

For individuals who cannot have a preference, however, the problem of how to assess whether existence is worth having must be decided by whether or not such a life has a favourable balance of satisfactions over miseries. It cannot be done by counting doomed fundamental interests because an individual may have a happy existence, an existence subjectively worth experiencing even though most basic interests have been doomed. But the point of asking such a question can only be to determine whether or not it would be right to bring into existence a person whose life would be like that. If the individual is already in existence the extent of the harm done her by being brought into existence in that condition is simply the extent of the disability. If life is so terrible for such a person that non-existence is clearly preferable then she should be killed. No moral person could stand by and see another creature suffer so much.

On the other hand, if the disability falls short of this high standard of misery, if I have been harmed by the dooming of one or more of my basic interests, then the damage has been done, whether or not as a consequence I would have preferred non-existence. If on balance existence is worth having even at such a cost then though I have been harmed I have not been wronged. On the other hand if it is not, I have been both harmed and wronged. Here again we see that the compelling way to characterise what we would want to say in such cases is the very opposite to Feinberg's formulation.

It is this consideration that leads Bonnie Steinbock to conclude, as we have seen:

> The escape from this dilemma is to see that it is not necessary to maintain that the child would be better off never having been born in order to claim that he or she had been wronged by birth. Instead we can say that it is a wrong to the child to be born with such serious handicaps that very many of its basic interests are doomed in advance . . . While this is something less than the right to be born a whole functional human being, it is not dependent on the implausible view that a life with serious impairments is always worse than no life at all.[31]

Steinbock wants the harms to be very serious indeed before she will allow that a child has been wronged by being brought to birth. So serious that they would constitute the dooming of some of the child's fundamental interests. Where this is the case she believes the child to have been wronged by life and thus to have the grounds for a successful wrongful life law suit.

Harming is not necessarily wronging

We must recall a preliminary conclusion, that it is simply wrong to choose deliberately to increase unnecessarily the amount of net suffering or harm in the world. It is not, however, a wrong against the individual thus created if that individual is or would on balance be pleased to be alive at such a cost. This is why we can, in appropriate circumstances, blame the mother without thinking that she has wronged her child who has in fact benefited by her actions although, and this is important, he has been harmed by what she has done. She has done harm, but she has wronged no-one.

Suppose a doctor was negligent in the sense of employing the wrong procedure given the diagnosis at which she had arrived but, because the diagnosis was in fact faulty, she had ended by benefiting the patient overall. Such a doctor has surely been negligent and as such might appropriately be disciplined by her profession. But, she has not harmed her patient. The patient has no grounds for compensation because he has not suffered by the negligence. Quite the reverse. The patient has perhaps been wronged because he has been placed at risk by being negligently treated, but there was no harm done.

Imagine a case in which there was no negligent medical practitioner, merely a mother. Suppose this mother has had a child she knew in advance would be disabled, but she reasonably believed that despite the disability the child would have a life worth living and if she wanted children, she had no alternative but to have such a child. Should her disabled child have a remedy in damages against her even though it has a life worth living and is glad to have been born? If there is injury, the mother has surely caused it, as much as, indeed more so than, any negligent doctor in a wrongful life suit. Indeed the issue of negligence is irrelevant. Suppose the child to have resulted from the mother's negligent use of contraceptives, does this make the resulting child more wronged?

It is difficult to believe that the mother has wronged her child. So far as her relations with the child she has engendered go she has benefited that child. It has a life worth living because of her choice. The idea that she might have an obligation to compensate her child for benefiting him is a nonsense.

In such circumstances wrongful life cases are simply misconceived. Not because the life in question has not been impaired, not because the individuals are not suffering, not because they have not been harmed; it has, they are, and they have: rather because it is not plausible to regard them as having been wronged.

You might harm someone in order to benefit them, but if so, you do not wrong them unless you violate their will in order to do so or breach some other obligation to them. The mother giving life with some measure of disability to a child who will find such a life worth having does not wrong her child. She is like the doctor giving a live-saving drug which has damaging side effects but side effects which are worth enduring for the sake of staying alive.

Should wrongful life suits succeed?

I think the answer to this question must be 'no' in almost all cases. First let's consider the case in which, while harmed, the resulting individual has a life worth living. The mother and/or the doctor may have harmed the child, but if that child has a life worth living, albeit a disabled one, then she has not been wronged by being brought to birth, although she has of course been thereby harmed. She was not been wronged by being harmed because, like those with Blighty wounds or those who have to endure the harmful side-effects of beneficial drugs, she has received a net benefit from what has happened to her and none of her rights has been violated.

It is for this reason that the present analysis of the difference between 'harm' and 'wrong' is more useful than that of Feinberg and others.

Secondly, we must consider the case in which the allegedly wronged individual has a life she finds worth not living. In this case she has been both harmed and wronged. And here I think she should have a legal remedy, at least if her condition results from negligence or some other wrongful act. But only for whatever satisfactions legal 'redress' might bring. It is better to ensure that the needy are compensated as of right rather than only if they have the resolution, patience, and perhaps also the resources to go to law.

There is, however, something paradoxical here. For if extra resources will make her life worth living then here life is only contingently worth not living and she will have been harmed but not wronged and fall into the earlier category. If, however, such resources will not make life worth living and if, as I have recommended, such resources as are needed will be provided as of right then there is no point in legal action except to grant the patient the death she desires and to inhibit other mothers and health professionals from bringing to birth individuals who will similarly have lives that are worth not living.

Rather than the unedifying spectacle of terribly disabled individuals suing well-meaning but misguided parents and health professionals, it would surely be better to make clear that to bring a child into being who will have a life so terrible that death is preferable is morally wrong. One way of signalling this fact in a social context is by legislation of course, but to punish individuals who transgress, whether by criminal action or action for civil wrong, seems merely gratuitously to increase human suffering. However, in the case of lives that are worth not living I concede that it might be unjustified to deprive the wronged individual of whatever satisfactions legal action and 'redress' might bring.[32]

In so far as we judge some people to be really better off dead, we should make it easy for them to achieve the death they seek by legalizing voluntary euthanasia. Where the disability is so great that they are incapable themselves of forming anything so sophisticated as a preference about life or death, and where again their life is so terrible that mere existence is a cruelty, then again we should give them a humane death by legalizing euthanasia in such cases.

Finally, if we think disabled children or adults should be compensated for their disability then we as a society should compensate them. Their need should be the trigger for the compensation, not the claim that their need results from wrongdoing.

In short, the problem of disability should be seen as one of social justice. In particular we should not distinguish between those disabled who are fortunate enough to have been wronged and therefore have a legal remedy available, and those whose disability is not a consequence of wrongdoing and consequently have no legal remedy nor a route to compensation.

5

Human Resources

One of the issues which our discussion so far has raised but not resolved is the question of the exploitation of human resources. By that of course I mean the resources represented by the human organism as a repository of therapeutic and research material and data. The question of the exploitation of data and information yielded by the study of human individuals, and in particular that yielded by the increased sophistication of genetic screening techniques and by the human genome project, we will postpone until Chapters 10 and 11.

Our present interest is in the immense possibilities for exploitation in every sense of that term, which the human organism at all stages of development represents. Biotechnology has sharpened our sense of the range of ways in which this might happen, although of course the issue of how far we are entitled to use one another as repositories of scarce resources is not a new one.

So far we have reviewed a number of possibilities. Biotechnology has made possible the establishment of human cell, tissue, and organ banks for transplant or implant purposes. The sources of such material might be human gametes, embryos, and fetuses whether alive or dead, *in utero*, *in vitro*, or in various stages of independence. Even the neonate or human adult might be, and indeed is already being, used as a source of such material. Animals of course are also a source of such material and some of the ramifications of the use of animals will be further considered in Chapter 7.

Exploitation

We are familiar with some aspects of this exploitation. The utility of human bodies as a source of blood and blood products has been exploited for many years; and organ, tissue, and other body products are now routinely transplanted from both living and dead donors. In

Norway it is now common for living donors to make up the shortfall in cadaver kidneys available for transplant, and as a result of this readiness to use live donors there is in Norway a balance between supply and demand in kidney donation. In the United Kingdom by contrast there is still an estimated shortfall of between 1,800 and 2,000 kidneys a year. In both the United Kingdom and Norway it has usually been accepted that good practice requires that donors be altruistic, and in particular that the sale of donor organs is regarded as unethical.

This problem of the commercialization of the supply of body products is now widely regarded as acute, and an eminent British surgeon has recently faced legal proceedings for his part in a 'kidneys for sale' scandal which involved the use of kidneys allegedly 'stolen' from an unconsenting Turkish citizen. The doctor concerned was subsequently 'struck off' the medical register for his part in the scandal.[1] The question naturally arises as to whether, if it is not wrong to give or 'donate' an organ or some tissue or blood, it can be wrong to sell the same products.

One of the focuses of such exploitation in connection with *in vitro* fertilization has been the issue of surrogacy,[2] but related problems arise in the case of sperm donation and egg or embryo transfer. In this chapter we shall be taking seriously the question: what constitutes exploitation and what is wrong with exploiting others? This sounds like a rather artificial question since it is usually believed that exploitation is simply and self-evidently wrong. However, as we shall see, this is a very problematic issue.

We shall try to make these problems manageable by considering the ethical issues under three broad headings. The first two headings will be considered in this chapter and to the third we will allocate a chapter of its own. First the question of exploitation of the dead, then exploitation of the living and finally, commercial exploitation.

In this chapter we will consider what is wrong with the non-commercial use of body products. This use is sometimes called 'altruistic' but I shall not be much interested in the motives of those who donate body products. I therefore prefer the more neutral expression: 'non-commercial'. In the next chapter we shall consider whether if it is not wrong to donate body products it can be wrong to buy or sell them.

I should make clear now that I am deliberately using and trading on the ambiguities in the term 'exploitation'. On the one hand it means simply 'making use of' some resource and it is contrasted with allowing that resource to go to waste. On the other it implies some form of cheating, of unfair advantage in the use of the resource. As the discussion develops we will see which of these two sense of the term applies to which cases.

I. Exploitation of the Dead

This is a large heading but the ethical problems associated with the use of cadaver tissue, cells, organs, and other products is, morally speaking, fairly straightforward. The dead, to adapt a phrase memorably used in another context,[3] are ex-persons, they are literally no more. They are beyond being harmed or benefited by us.

The interests of the dead

There is of course a legitimate though somewhat artificial sense in which we may talk of the dead having interests which persist. I mention it for the sake of completeness though its plausibility or otherwise plays no role in the argument we will be developing. It does seem natural to say, however, that I have an interest, for example, in the continued existence of this planet, and this interest is not extinguished by my death, for my interest in the planet's continuation is impersonal. Equally and perhaps more clearly, I have an interest in the survival of the tropical rain forest or in my own reputation remaining untarnished and so on, and I have such interests even if I never see a rain forest on the one hand, or know that I have been slandered on the other. And the interest I have in both these things persists and survives my death, for my reputation may still be damaged and it remains true for as long as I have such a thing as a reputation that losing it is not in my interests. But while such persisting interests may be damaged, the people whose interests they are cannot be harmed once they are dead.

It is, I admit, hard work imagining why one should separate harming someone's interests and harming that someone. But the point of doing so is perhaps this: if we damage the environment irreparably today, this will harm the interests of future generations but it will not harm individuals as yet undifferentiated until they come into being. It harms their interests now and them when they exist. Similarly the interests of actual people persist after their deaths. When they are alive you can harm (or benefit of course) both the individual and her interests. Once she is dead only her interests remain to be harmed.

This is why the damage to the persisting interests of the dead must be set against the damage to the persisting interests of the living, damage which, in the case of the living, also affects the persons whose interests they are. This double damage will for all practical purposes always give the edge to the interests of the living.

The effect on the living

The only other thing wrong with using cadaver products is the effect on living people. On the one hand there is the benefit to be derived from using the products; whether this benefit is in terms of saving the lives of existing individuals, or for other therapeutic uses, or whether the proposed use is research where the benefits are future and more speculative. On the other there is the possible harm to those who do not wish their own body products to be used after their deaths or to those who do not wish the bodily products of their relatives or friends to be so used.

We saw in Chapter 2, that 'sentimental morality' is really no morality at all. And a *feeling* that one's own organs or those of loved ones should not be used, or that our bodies should not be *desecrated* after death is not necessarily a moral feeling. But even if such feelings were to be given moral weight, they would have to be balanced against the powerful moral reasons for using bodily products in contravention of those feelings. If we can save or prolong the lives of living people and can only do so at the expense of the sensibilities of others, it seems clear to me that we should. For the alternative involves the equivalent of sacrificing people's lives so that others will simply *feel* better or not feel so bad, and this seems nothing short of outrageous. Although *this* is not of course the reason or even a reason against it.

Of course, where, as at present, most societies have a voluntary system for donation of bodily products it is important not to alienate the potential donors, or frighten them off altogether. Equally it is important to be sensitive to the sensibilities of those, relatives perhaps, whose permission will be necessary if body products are to be made available to therapy or research.

It is widely agreed that if the permission of the relatives of the deceased is necessary, then the deathbed is neither the most considerate, nor the most opportune, place to ask for it. Nor is it exactly tactful to ask a dying individual if he wouldn't mind parting with those parts of himself that will be surplus to requirements in the near future, that is of course with every bit of himself.

The obligation to use bodily products

But while it might not be sensitive to ask, it would still be wrong of the relatives, or indeed of a moribund individual, to refuse. It would be wrong because people cannot plausibly claim that their sensibilities are more important than the lives or the health of other present or future people.

If you doubt this claim just ask whether someone would be entitled to cause the death of another merely because they found him offensive or because this other individual had otherwise damaged their sensibilities, or where that individual's continued existence would damage their sensibilities.[4] These would be poor justifications for murder or manslaughter and so they are in the context of cadaver donations also. For if it is clear, as it is, that for want of an organ, or some bone marrow, an individual will die, then the failure to give those bodily products or permit them to be given causes death as sure as shooting.[5]

Cadavers should be public property

The solution to the problem of sensibilities is of course to determine that cadavers, like the foreshore, belong to the state and that therefore neither relatives nor the former 'owners' of the cadavers would have any binding interest in their fate. People would, I believe, soon get used to the idea, particularly if there was a concerted campaign of education and argument, and the automatic public ownership of dead bodies and their bodily products would remove the need to interpose intrusive requests between people and their grief.

Indeed it seems clear that the benefits from cadaver transplants are so great, and the reasons for objecting so transparently selfish or superstitious, that we should remove altogether the habit of seeking the consent of either the deceased or relatives. This we already do when post-mortem examinations are ordered without any consents being required and despite the fact that these too involve interference with the dignity of a dead body and the removal (albeit temporarily) of organs. It has always seemed to me curious that the state can order a post-mortem examination to satisfy its curiosity about the cause of death, but not order cadaver transplants in order to save the lives of living citizens. Of course, post-mortems are not usually ordered out of simply curiosity, there are public safety and public policy considerations. It is important that the cause of death be known in case the same cause represents a further danger to the community, whether that danger be in the form of a disease or contagion, or in the form of a possible murderer at large. But again related but more powerful considerations weigh in favour of public ownership of cadavers.

The definition of death

Now while the automatic public ownership of cadavers and cadaver products is, I believe, a truly simple approach to the problem of the supply of cadaver organs and other parts, it will not go very far to solving the problem. For one thing most of the parts required for transplant and other related purposes must be taken from living bodies. Some of these living bodies it is true will be 'brain-dead' or 'brain-stem dead' or 'whole-brain dead', but these are *not* definitions of death, although they are usually offered as such,[6] rather they are suggestions as to when it is legitimate to seize bodily parts. For living bodies with 'dead' brains, whether these brains are dead in whole or in part, are just not 'dead' in the common understanding of this term. There will always be a real and defensible sense in which they are still felt to be 'alive' They are not stiff and cold, they are not 'lifeless'. Of course they can be declared to be 'dead' for legal, transplant, inheritance, or any other purposes you care to invent, but this will be a convenient fiction. The real ethical issue is simply the question of whether or not such individuals have retained any moral importance, or 'value'; in the terminology I prefer, it is the question of whether or not these individuals are still 'persons'. What they can definitively and finally be declared to be is ex-persons or non-persons, or, what amounts to the same thing, no longer of a moral importance remotely comparable with that of those they can be used to help.[7]

But perhaps I run too far ahead. Let us backtrack and consider the second broad question, that of the legitimacy of the exploitation of the living.

II. Exploitation of the Living

This will prove to be by far the most important category of exploitation and it raises the most acute problems. Here we must again distinguish at once between two categories of the living: those living persons who may provide voluntarily or involuntarily, altruistically or commercially, their bodily products for transplant or use by others; and those living human individuals who are not and will never be moral beings, and hence are not and will never be persons, and the question of whose consent to donation does not therefore arise.

We will deal first with the question of bodily parts from living human individuals who are not persons and will never be persons.[8]

Anencephalics

Perhaps the first and clearest case of a human individual who is not and will never be a person is the anencephalic infant, an infant born with most or all of its brain missing. Such infants will never become beings capable of valuing or experiencing a life. They may, it is true, live for a few days or even weeks supported by their own assisted breathing and cardiovascular system, if they are given food, fluids, and antibiotics. Although most such infants are 'brain-stem dead' some may have some electrical activity in the brain-stem.

These anencephalic infants have no possibility of becoming self-conscious beings, persons properly so called. The use of their organs seems to me to be entirely analogous to the use of 'cadaver' organs, that is organs from bodies kept alive on life support machinery although they have permanently lost all that makes life worth having, all that makes them a person.

If we take next the possibility of using the embryo or indeed the aborted fetus as a source of bodily products we can see at once the full range of issues involved.

Banks of embryo cells and tissue

Embryos might be used not only as a source of living cells, tissue, and perhaps even organs for transplant purposes, but might be used to create 'banks' of cryopreserved bodily products for future transplant. There are a number of different sorts of possibility here which we reviewed in Chapter 1. Now is the time to try to assess the ethics of implementing such possibilities. We must first remind ourselves of what is, and might become, possible.

Very early embryos have cells which, as we have seen, are toti potential, that is to say they are at a stage of development which leaves them the potential to turn into any part of the body or indeed into extra-embryonic tissue. These toti-potential cells also have the property of being easily accepted by an alien immune system, they are not 'recognized' as alien and so are not rejected by a genetically unrelated host. This naturally, as we have seen, makes them highly eligible for transplant purposes.

Organs and other material may be transplanted successfully not only from embryo to embryo but from embryo to adult. This has proved very successful in animals and there is every reason to suppose this will also prove to be true in the case of humans.[9]

Cloning may also offer transplant possibilities at later stages of development. A human embryo may be 'cloned' by division *in vitro*, resulting in twins or quads. Some of the resulting individuals may be kept as a 'reservoir of antigenically identical spare parts for later life',[10] although as Mark Ferguson has pointed out, 'research into the control of antigenic determinants on embryonic cells would render this complex procedure unnecessary from a therapeutic standpoint'.[11]

If a different cloning technique became routinely successful it might be possible for an adult to create an embryo which was a clone of herself which, again technology permitting, might be grown to a later stage of development to provide a source of organs and other bodily products for transplant. The process likely to be used would be to take the nuclei of adult cells from the chosen adult beneficiary and substitute them for the nuclei of an early human embryo and then, being a clone, it would share the genetic make-up of the adult nuclei donor. As we mentioned earlier,[12] this technique is currently extremely difficult and therefore problematic as a routine solution to transplant problems. However, as in so many areas of biotechnology, it would be foolish to assume that this will remain the case.

All these possibilities have caused marked feelings of unease and sometimes horror. But are such feelings justified? Some commentators have attempted to deflect adverse reactions by concentrating on the idea that only fetuses that are surplus to requirements and whose bodily products would otherwise go to waste will be used.[13] Clearly where this is the case, there is a powerful argument for using such bodily products for it cannot be morally better to do nothing than to do something good. It cannot be morally preferable to waste a resource, *any* resource, than to use it for a beneficial purpose, whether that purpose is a therapeutic or a research purpose. But we should not assume that only such spare embryos will be used or that there will be sufficient numbers of such embryos to meet the likely demand. The question then arises: should we supply the demand for embryonic tissue cells and organs by specially creating embryos as transplant donors?

Transplant embryos

As we have noted, Robert Edwards distinguishes between 'spare' and 'research' embryos. Spare embryos are those which arise as by-products of *in vitro* fertilization programmes and become spare when their donors have no further use for them. Research embryos are, on the other hand, produced purely for research purposes.[14] Research embryos are usually obtained by asking women who require sterilization whether they would be prepared to donate research embryos prior to sterilization. This process is not without its risks as there are some problems associated with the hormone stimulation of the ovaries.

As we have seen, Edwards thinks it only justifiable to use spare embryos for research, he mistrusts the intent of scientists who would 'create such embryos purely for research'.[15] So long as there are sufficient spare embryos for research and transplant purposes I can see the force of Edwards's distinction, though not for his reasons. My acceptance of the moral force of the distinction stems from the arguments that I can see for not asking women to run any extra risks which will be of no benefit to themselves. But suppose sufficient spare embryos do not arise as by-products of *in vitro* fertilization?

Should we deliberately create embryos for transplant?

How morally important is having a baby? We have already looked at some aspects of this question in Chapters 3 and 4,[16] but the question for us now is this: if it is justifiable to produce spare embryos as by-products of the quest for a live birth, could it be wrong to produce them as means to save a live *life-in-being*? If the objective of having a *new* baby is one that can legitimately produce spare embryos, is it so implausible that the objective of keeping alive an 'old baby' might not also make comparable moral demands?

This is an important point and it is worth labouring because it forces us to concentrate on the question of what precisely does justify the creation and 'sacrifice' of an embryo? It seems to come to this: either the mother is entitled to create her own embryos and do with them what she will, or she must justify their creation in moral terms. If the former then there is no question to answer. If the latter, we are entitled to concentrate on the magnitude of the moral justification offered. If it is acceptable to produce spare embryos in pursuit of successful pregnancy then it must be justifiable to produce them in pursuit of something plausibly of the same moral magnitude. Saving a life-in-being surely comes into this category.

There is of course a further moral issue here. It is the question of whether it is reasonable to ask one person to take risks to save the life of another. This is a real and difficult question and I will not address it directly here, for its answer turns on a balance between risk and pain to the rescuer on the one hand and certainty of danger to the person whose life is at risk. It is worth noting in passing that there do seem to be clear cases where there is such an obligation—to give blood for example, or maybe bone marrow, where the risks and inconvenience or pain are small and the benefits clear. However, if someone is *willing* to take the risk after its nature and scope has been carefully explained, then it is difficult to see why these embryos should not be used to save life as much as to create it.

We must now turn to one or two further objections that might be made, and since they have been recently invoked in consideration of the question of using material from aborted fetuses we will consider them in that context. But first a brief digression.

Digression: the argument from myopia

I have just suggested that if someone is willing to take the risk, 'it is difficult to see why' embryos should not be created for research. Of course nothing follows from the fact that I am myopic to something, nor indeed from the fact that most people might be similarly unable to see it. The argument from myopia as I call it, has no force at all, although this fact has done little to reduce its popularity. It is, however, a tempting locution, and I have used it at a number of points in this book. Always, I hope, to make a legitimate challenge and not a spurious objection. The spurious objection is of course that *my* failure to see something is *no evidence at all*[17] as to its not being there to be seen. The legitimate challenge, and the sense I hope in which I have used this term, is simply to record that no plausible arguments or evidence have been produced to show why whatever it is should not happen, and that in the absence of such arguments or evidence 'it is difficult to see' an objection. The use of the phrase is simply a claim that there is an initial plausibility to a particular position and that no plausible argument to the contrary has been produced, and a challenge to produce such argument.[18]

The aborted fetus as transplant donor

A feeling of unease sometimes attends the use of material from aborted fetuses. Reports that such material will be used to treat Parkinson's disease, for example, have met with opposition.[19] This seems, however, to be a straightforward example of cadaver transplants. It is true that the fetus cannot consent to its cadaver being used for transplants, but then there is no question of its withholding consent either. It is not the sort of being whose own consent can be relevant one way or the other, for it is not capable of forming a view about the matter.[20] So long as the mother of the fetus consents, then the situation is entirely parallel with other cadaver transplants, from children for example, and there seem to be no peculiar ethical issues which need special consideration. This, however, has not prevented people from inventing them. Let's look at some.

Melanie Phillips writing in the *Guardian* newspaper[21] has suggested a number, of which we shall consider two.[22] She claims that the knowledge that fetuses can be used to help someone with a dreadful disease will be an inducement to abortion and that it is repugnant to use aborted tissue even for a noble purpose because this would be to adopt the (supposedly dubious) principle that the end justifies the means.

The first argument, to the effect that a mother's wish to help a sick person might induce her to undergo an abortion, is a complex one and one that has been taken up by a number of writers on this issue.[23] It is surely doubtful that a mother who wanted her child would be much persuaded by such an argument, and a mother who does not want her child does not usually need a further incentive to an abortion since she can almost always obtain her desired abortion with relative ease. It is of course possible that a woman who has decided on an abortion for other reasons might find some consolation in the possibility of some good or extra good coming of her decision, and this is surely to be welcomed.

However, it is possible that women who do not want children or further children of their own might be induced to become pregnant with a view to abortion in order to help someone, particularly perhaps if that someone is known to them or if there is some financial inducement. I will leave the discussion of the relevance of commercial considerations until the next chapter and for the moment confine myself to the question of the ethics of someone deliberately and for altruistic motives becoming pregnant in order to generate tissue for transplantation.

Good reasons, good deeds, and good people

Let me start by relating a real[24] case history: A woman, we shall call her Martha, sought medical assistance with establishing a pregnancy via artificial insemination by donation from her estranged husband. She had one daughter already who suffered from a form of leukaemia which required a bone marrow transplant. There seemed no prospect of her finding a compatible bone marrow donor for her daughter and so she wished to have another child by the same father in the hope of providing a compatible sibling who could donate the required bone marrow. Since for any given sibling of her daughter there would only be a one in four chance of the required compatibility, the mother's intention was to have any pregnancy tested for compatibility *in utero* with a view to abortion if the child was not of the required genetic constitution. The mother intended to persevere with her AID from the estranged husband until she achieved a compatible donor sibling. This second child she intended to love and cherish but she would consent on its behalf to the bone marrow donation. For the donor, bone marrow donation is an invasive procedure requiring drawing off bone marrow from the hip but it is not dangerous and the pain, I am told, is equivalent to a hard kick on the thigh with consequent bruising and discomfort lasting for a few days.

The dilemma as perceived by the medical staff to whom this proposition was put was: should they help to establish the pregnancy via artificial insemination, should they carry out the genetic screening and arrange the consequent abortion if necessary, and should they carry out the bone marrow transplant on a child bred primarily for that purpose?

We will consider the ethics of this mother's choice in conjunction with those of the analogous case in which a woman contemplates becoming pregnant in order to have an abortion to provide tissue for transplant purposes.

Good people and good deeds

There are two pairs of related questions here. One is: what is a good reason to have an abortion and what justifies abortion? The second pair of questions is: what is a good reason to have a child and what justifies having a child? The answer to the first question in each pair will tell us something about the moral character of the person making the choice, but not necessarily anything about the legitimacy of that choice. The second question in each pair is directed to the issue of whether or not the person is entitled to pursue her choice. This distinction reveals an important difference between what is involved in the moral assessment of persons and what is involved in the moral assessment of actions—good people can do bad things, bad people can do good things, good things may be done for bad reasons and vice versa. This is commonplace of course, but it is a distinction that is often lost.

Let's take the second question in each pair first—what justifies an abortion and what justifies having a child? If the argument of Chapters 2 and 3 is right, then a woman is entitled to abort a fetus, and justified in so doing if she chooses, because the moral status of the fetus is not such as to protect its life.[25] In Chapter 3 we saw that while the decision to bring a child into the world is not unproblematic, if the child is expected to be healthy and perhaps happy, then this is certainly not only a justifiable but also a good decision.[26] If these answers to the second question in each pair are right then parents do not need specially to justify having a child who will be healthy and probably happy; and, on the other hand, an abortion to generate material for transplant is justified because the abortion is justified.

Martha's choice

If we turn to the question of the legitimacy of Martha's choice to have an abortion if the fetus is not of the 'right' genetic constitution, the situation is the same. Now let's consider Martha's choice to have a child and to use that child as a bone marrow donor for its sibling.

Suppose Martha already had two daughters, only one of whom needed the bone marrow transplant and the other was a suitable donor. Let's call this case *Martha* 2. We'll assume both are too young to give an informed consent themselves. Would the mother be wrong to choose to donate the bone marrow of one sibling to the other?

This is a complicated but ethically fairly straightforward question. It raises of course the question of the legitimacy of subjecting those who cannot consent to procedures that are not of benefit to themselves. Many writers on health care ethics have made very heavy weather of this issue and have followed various codes of practice in holding that it is never right to subject someone who cannot exercise choice, particularly a child, to a procedure that is not therapeutic or at least beneficial[27] for that child, for the subject of the procedure. There are many areas of life in which we recognise that it is reasonable to subject people to some risks and possibly to some degree of pain and distress for the benefit of others. A useful reminder here is:

The ship in distress. If a ships captain diverts his passenger ship into a storm to go to the aid of a stricken vessel, he puts his passengers at some extra risk to their lives and certainly will knowingly subject them to anxiety, distress, and even pain and injury. The risks will of course vary, but they may be substantial. If the seas are very rough, minor injuries from falls and so on may be almost inevitable. His justification is surely that he must weigh against the extra risks to his passengers the fact that others will certainly lose their lives if they are not rescued. It is generally accepted that captains of ships have both a right and a duty to act in this way.

The appropriate test to apply here seems to be to set losses against gains. Against such a calculus the mother seems well justified in subjecting one child to a small degree of pain and risk for the sake of the life of her other child.

Now would Martha be wrong to create a child with a view to using her as a bone marrow donor? We saw in Chapter 4 that the child herself could not claim to have been wronged, for this would be a price well worth paying for the chance of existence (*ex hypothesi*, that child's *only* chance of existence). True the child would have been harmed slightly since pain is a harm, but again a harm well worth trading off against the chance of existence and indeed, as we have just suggested, one which that child ought to be prepared to pay to save the life of another child, whether a sibling or not; and, moreover, one which the child's parent is entitled to make on her behalf. Ask the question: could the mother be right to deny one child her life rather than subject the other to a minor amount of pain and distress and a very small risk to life or health?

We have considered the questions of the use of embryos and fetuses as transplant donors and also the case of young children as donors of bone marrow which, like blood donations, will be replaced by the natural operation of the donor's body. But what about the donation of irreplaceable organs and tissue, like kidney donation for example?

Live organ donation

If we start by trying to get clear about the ethics of voluntary, non-voluntary, and even involuntary organ donation on an 'altruistic', that is a non-commercial, basis we will be better placed to see what difference commercial considerations make.

An immediate and important difference between organ donation and most of the cases we have so far considered is the increased level of risk involved. This risk is of various kinds. There is the risk of the transplant procedure itself. There is always a risk involved with general anaesthetic[28] and there are the added risks of major surgery including post-operative dangers. Finally there is of course the risk to an organ donor that she might subsequently need the missing organ. If a living kidney donor subsequently experiences kidney failure she may need a kidney transplant herself.

It should be emphasized, however, that it is generally accepted that the risk to the organ donor is small (about 1/1,600 risk of surgical mortality and 1.8% risk of major complications) and if she survives the operation successfully it is accepted that her risk is *no* greater than that of a normal person who shares her other characteristics.[29]

The feedback effect

We should note one important feature of organ donation which reduces the risks of being a donor. This is the feedback effect which occurs to the degree to which live organ donation becomes widespread. The more chance you have of receiving an organ donation, the less risk there is when one of your organs fails. So that the danger of, say, your one remaining kidney collapsing after donating your first kidney to a needy recipient is reduced in proportion to the probability of your receiving a successful transplant to replace the failing organ. Your readiness to donate a kidney, if it is part of a general pattern of such readiness, feeds back into the whole system of organ donation in a way which reduces the risk you run when you donate your organ. But again, since this risk is vanishingly small to survivors of the operation, the feedback effect while comforting is not particularly important.

Voluntary donation

We are entitled freely to choose to run risks. If I decide that I would like to donate one of my kidneys and run the risks of the procedure and the risk that I might subsequently have kidney failure, then it seems that this is a matter for me. Like all the other risks that I choose to run, from driving a car to enjoying wine or playing squash, these are matters of personal choice. Of course this acceptance of the legitimacy, indeed the primacy, of autonomous choice presupposes that the choice is autonomous—freely made, uncoerced, and so on.

Should I be permitted voluntarily to donate a vital organ like the heart? Again, if I know what I am doing then I do not see why I should not give my life to save that of another if that is what I want to do with my life. Non-voluntary or involuntary donation of vital organs is of course murder and is justified only where murder is justified.

Non-voluntary donation

Non-voluntary donation occurs when an individual who cannot give a valid consent is 'volunteered' by another as a donor. I will confine discussion here to the cases of individuals who are or will be persons, that is individuals who are capable of valuing existence and of having 'a life'. This might happen in the cases of embryos and fetuses who are intended to be born alive and permitted to grow to maturity,[30] or in the case of children who are too young[31] to give consent, or in the case of mental patients.

If we changed Martha's case so that the needed transplant item was a kidney instead of a bone marrow we would have one suitable example here. Another occurred in the United States case of *Strunk* v. *Strunk*.[32] In this case a young married man, dying of kidney disease, asked the court for permission to use his younger brother as the live kidney donor. His brother was a 27-year-old mental patient, confined in an institution, who was reported to have a mental age of 6.[33] In this case the court used an antiquated principle of English law, the doctrine of substituted judgement, which in the case of lunatics, allows the courts literally to 'substitute' their own judgement for that of the lunatic where they believe that the substituted judgement is the judgement the lunatic would have made had he been in possession of all his faculties. They reasoned that because the younger brother was allegedly psychologically dependent on his older brother, his well-being would be damaged more by the loss of his brother than by the loss of his kidney.

It seems far from clear that this is really the judgement that Strunk junior would have come to if he had been in possession of all his faculties. But if 'substituted judgement' is a doubtful principle to adopt in the case of live organ donations from children and mental patients, what principle should we adopt?

Risk assessment

A crucial feature here is of course the degree of risk involved to the donor. In the case of bone-marrow transplants the risk was relatively small[34] and the advantage great. With something like live kidney donation the advantage to the recipient is still as great but the risk, even taking account of the feedback effect, is greater also. While as we have noted the risk is still relatively small it is none the less significant and one which no reflective person would run lightly. The question then is what justifies someone imposing this risk on another who cannot consent?

Here again we should start at any rate by distinguishing the justification that might be open to a parent, or someone like the Court acting *in loco parentis*, and a general justification.

Would it be right for a parent to use the kidney of one of her children to save the life of another? Here the considerations seem very near to those of *Martha 2*. Let's call this case *Martha 3*. If it would be wrong for Martha 2 to let one of her children die rather than subject the other to a relatively small risk and some pain and inconvenience then why would it not be wrong for Martha 3? After all, so the argument might go, there is only a small increase in the risk to the donor child of Martha 3 compared with that facing the donor child of Martha 2. It would seem invidious to sacrifice the life of Martha 3's child while letting that of Martha 2 stay alive.

The argument from inconsequential differences

This is of course *the argument from inconsequential differences* and it can be used to prove a number of impossible conclusions; for example that you can go on adding grains of sand one to another and never produce a heap of sand. This is because, so the argument goes, starting from just one grain and adding grains one by one, the addition of something as small as just one grain could never make enough of a difference for a heap to appear where there was formerly no heap. And yet of course we know that this process inexorably produces a heap of sand.

So, we have a reason to be suspicious of an argument that uses inconsequential differences, that argues because one case is only a little further along a continuum it cannot have reached a point where a difference in quality rather than a difference merely in degree has occurred. Even inconsequential differences added together can make for a *consequential difference*, a difference that makes a difference! How can we find a way of thinking about what Martha 3 should do?

Suppose a mother was fleeing with her two children from a cruel and murderous enemy, perhaps a Jewish mother fleeing with her two daughters from the Nazis. Suppose one daughter was ill and so an impediment to fast movement. The mother knew she could get clean away if she abandoned the sick daughter, but, by insisting on taking both, she placed the healthy daughter at a slight risk and caused her pain comparable to the kidney transplant operation. Say, for reasons we will not elaborate because they would be too fanciful, she knew that the healthy daughter would be shot during the escape but would probably make a complete recovery. That is, would have the same probability of making a complete recovery as if she had donated a kidney and would undergo relatively comparable pain. We can put a conservative figure on this. If the risks to the kidney donor are of the order of 1/1,600 risk of death, would the mother be wrong to save her sick daughter at this risk to her healthy daughter? Would she be right to abandon her sick daughter to certain death rather than subject her healthy daughter to such a risk?[35]

Here I think we would judge the mother to have made the right choice if she tried to save both. But even those who doubt this would surely think she had made a defensible choice—a choice she was not wicked (unethical)[36] to have made, and one that she should not at any rate have been prevented from making.

But now suppose that the two children she was protecting from the Nazis were not her own but simply two children she happened to have met along the way. Should this alter her decision if she were to be faced with the same dilemma? Surely not, because she has a duty to both children. If she must not expose one to avoidable pain and risk, she must not expose the other to avoidable death.

Keeping it in the family

We can see now that it is ethically very difficult to keep it in the family. For if all children matter equally, then surely we cannot choose automatically in favour of the one whose danger is least! If your child is suffering from fatal kidney disease and mine has the only available kidney, would I be wrong to risk my child to save yours? I do not think I would in fact do so, but that is because I am a parent and I have fierce protective feelings towards *my* child. But would I be morally wrong to risk my child to save yours? And of course there is a political dimension to this also. Would the state be morally at fault if it were to encourage a system of live transplants across society whether adult to adult or child to child? Of course the adults could give valid consents and so this would present no problem. But if it was not wrong and moreover a good thing for one adult to do for another, would it be wrong for a parent to consent on behalf of her child?

I suspect that one reason that would be invoked against such a practice would simply be that the child would in fact be unlikely to consent and that therefore any claim to 'substituted judgement' would be spurious. But is this only a contingent objection, an objection contingent upon the fact that altruistic donations between adult strangers are rare and are likely to remain rare? If altruistic live transplants were common, and this sort of practice was widely accepted as right, then the judgement that the child would be likely to object if capable of being asked would be suspect.

If we came across a society which operated along these lines and accepted as right and proper the voluntary live donation of organs between members, whether children or not, it is not clear to me that we would judge or that we ought to judge such a society as morally worse than our own. It might of course be rather difficult to move our society in this direction and whether or not this would be desirable might depend upon whether or not we could provide enough donor parts in other ways. But with genetic manipulation making the chances of successful transplantation better, these issues are likely to remain important.

I do not wish to pursue these questions further here. I have examined them at length elsewhere[37] and in any event they take us too far beyond our present concerns. For the present we can safely conclude that non-voluntary transplants between children under the exclusive protection of the same adult, whether parent or not, are defensible if they offer the best chance of both children surviving in reasonable health. Whether this reasoning extends to a defence of the decision in *Strunk* v. *Strunk*, I leave to the reader's good judgement. We must now turn to an examination of the difference that commercial considerations make to all of these issues.

6

Commercial Exploitation

The question of the commercial exploitation of the human body and its component parts is vexed in the extreme. A possible opening gambit might be to begin with an analysis of the concept of property and discuss whether 'ownership' is a relationship between persons and objects or whether it is rather a question of rights, whether legal or moral. On this latter view, perhaps the most plausible, property is simply the right to exclusive use and disposal of things, whether 'material' like organs, or 'incorporeal' like ideas.[1] I will not, however, attempt any discussion of the *nature* of property or ownership. Instead we will be interested in the ethical and practical question of what should be done with the bodily products of individuals at various stages of development, and in the question of who should control this disposal. In particular we shall consider the issue of whether or not anyone should be paid for parting with their bodily products or for consenting to the removal of parts from others and with the related question of whether or not it might be wrong to purchase such products.

Since we have already dealt with many of the issues concerning the ethics of the use and disposal of bodily products the main question remaining is:

> If it is right to give something away, can it be wrong to sell it?

Again there are three separate dimensions to this problem. On the one hand there is the case of tissue, cells, and organs from the gametes, the embryo, and the fetus. Then there is the problem of children, and finally there is the problem of whether adults may freely sell their bodily products.

From gametes to fetus

If we start by considering the legitimacy of trading in gametes, or embryos, the most important issue is simply whether anyone should have the right to exact a price for the use of such things as gametes, embryos, and fetuses. I have suggested that cadavers, or the moral equivalent of cadavers, that is living human non-persons who will either never be persons or will never again be persons, should be publicly 'owned'. That is, that their parts should be *freely* available for beneficial use. Does the logic of this suggestion extend to gametes, embryos, and fetuses?

This is a complex question because cadavers come into existence willy-nilly, people do not usually choose whether or not to die or become ex-persons, and there is usually no point in offering incentives to individuals to become cadavers. This is not always true and we will look in a moment at the question of whether it might be legitimate to buy and sell vital organs like the heart in the case of living 'donors'. However, while the offer of an opportunity to die prematurely for the benefit of others is likely to have limited appeal, people do frequently choose to make available gametes and embryos.

Where gametes are in demand, different practices have been evolved depending on the gender of the gametes. Where male gametes are concerned, as in Artificial Insemination by Donor, (AID), it has been customary to pay donors a small sum for their trouble. For the provision of female gametes, eggs, on the other hand, it has been customary to ask women if they would be prepared to donate eggs free of charge.[2]

It is an interesting question as to whether this difference constitutes unfair discrimination against women and a further example of sexism, in denying them fair payment for provision of an important resource. On the other hand it might be seen to be expressive of the greater respect shown to women. If it is thought to be expressive of greater respect, this might perhaps be because it also demonstrates a disinclination to subject women to what might be considered to be a form of prostitution involved in the sale of semen, in that men are induced to masturbate and sell their semen for money. I shall not try to resolve this question here.

Compensation or inducement?

Would it be wrong to provide financial compensation to women to provide eggs for therapeutic purposes? Would it be wrong to provide financial inducements? Here there seems to be an important difference in 'feel' between *compensation* and *inducement*. There are two possible sources for this difference. On the one hand 'compensation' seems to imply recompense for things past and it does not *sound* as though something like compensation would influence events. What has happened has happened for its own reasons and the compensation comes after the event. 'Inducement', on the other hand implies that the inducement figures as one of the operative reasons for the event's taking place. This distinction is almost certainly illusory in this context, for if compensation for the provision of gametes were to become established practice, then it would be known in advance that compensation would be available and in so far as the levels of compensation were sufficient they might inevitably act as an inducement.

Here the most significant issue would be the level of remuneration. But if the level were sufficient to compensate it might well also be sufficient to induce.

But what is wrong with inducement? To answer this question we must consider the concept of exploitation. We will do so now because our understanding of whether or not the existence of remuneration constitutes exploitation will help us to consider the ethics of remunerating children and adults also.

I. Exploitation

There are two dimensions to the concept of exploitation, or rather, as I would prefer to describe it, two differing conceptions of the same concept of exploitation. The one involves the idea of *wrongful* use and may occur when there are no financial or commercial dimensions to the transaction. A classic case here would be where it is claimed that lovers may exploit one another, that is, use one another in some wrongful way.[3] The most familiar of such wrongful ways in this context might be where it is claimed that one partner uses the other or treats the other merely as a 'sex object'.

The second conception involves the idea of some disparity in the value of an exchange of goods or services. Sometimes both elements or conceptions are involved and the claim that there is exploitation is complicated and even confused.

This characterization of exploitation is echoed by Joel Feinberg in one of the most comprehensive contemporary discussions of the concept of exploitation. Feinberg asks:

What is the difference between one person merely 'utilizing' another for his own gain, and one person exploiting the other? The correct short answer to this question, of course, is that there is an element of wrongfulness in exploitation that distinguishes it from nonexploitative utilization.[4]

1. Exploitation as wrongful use

It is often claimed that the existence of financial interests is a sure sign of exploitation, indeed is what might turn utilization into wrongful use and hence into exploitation proper. This passage from the Warnock Report[5] is not untypical:[6]

> That people should treat others as a means to their own ends, however desirable the consequences, must always be liable to moral objection. Such treatment of one person by another becomes positively exploitative when financial interests are involved.

This was produced by Warnock as an objection to the practice of surrogate motherhood, but if it has any validity it will necessarily apply even more strongly to the cases we are considering. The idea here seems to be that it is the existence of the financial interests that is the feature of the situation that makes the wrongful use exploitative. As we shall see, however, Warnock must be making a hidden assumption about the wrongfulness of the practice *per se* or about the inevitability of coercion.

Protecting the vulnerable

Another approach to exploitation has more self-consciously moralistic motives. R. E. Goodin starts with the assumption that certain uses of people are exploitative because they violate certain intuitive judgements. Interestingly Goodin picks on the same example as Warnock, that of surrogacy, and like Warnock, Goodin's moral intuitions seem indistinguishable from moralistic prejudices. 'Someone willing to sell a cornea or rent a womb is (or soon may be) legally prohibited from doing so. Presumably it would be wrong—exploitative—to use a person in such ways, even with that person's complete consent'.[7]

It is true that Goodin has a theory which underpins his prejudice, but it is difficult to tell whether it is the prejudice that is driving the theory or vice versa. He believes that the common thread uniting all cases of exploitation is that they violate a principle requiring protection of the vulnerable.[8] However, it is not clear that a surrogate mother is necessarily vulnerable or more vulnerable than the distressed couple to whom she may 'rent her womb'.[9] This is not the place for a full analysis of Goodin's ideas but it is worth noting that the injunction 'protect the vulnerable' may not relevantly distinguish the various parties to a case of possible exploitation. Goodin tells us that

> The duty to protect the vulnerable is, first and foremost, a duty laid upon each and every one of us not to do anything which would constitute taking unfair advantage of those who are peculiarly sensitive to our actions and choices. That is to say, the duty to protect the vulnerable gives rise, first and foremost, to a duty not yourself to exploit those who are vulnerable.[10]

But to take his own case, that of surrogacy, even if we grant that the womb lessor is vulnerable, so are the would-be lessees, they desperately want the child that only the surrogate can give them. If we try to choose between the two by trying to decide which of the two parties is *the more vulnerable* Goodin's principle would be violated, for inevitably one vulnerable party would be exploited. If, on the other hand, we outlaw the practice as a whole, we may act to the detriment and violate the wishes of both sets of vulnerable individuals! It is a puzzle how such a course could constitute caring for the vulnerable even though it might just amount to avoiding exploiting them.

The moral question it seems to me is: what moral costs attach to the sort of 'clean hands' policy advocated by Goodin and, is it not perhaps better, to have dirty hands if keeping our own hands clean is so damaging to others?[11]

For many, a further problem with Goodin's principle would be its unashamedly paternalistic character, involving as it does the prevention of people implementing what he concedes may be completely free self-regarding choices. In any event, in many real cases, it might well operate to prevent the vulnerable from utilizing the only assets available to them. It might be that we ought, on public policy grounds, to do this, and we will look again at this problem later in this chapter.

Who is most vulnerable?

It seems so often to be assumed that those who may be exploited are the only vulnerable parties to a possibly exploitative transaction. However, it is important to re-emphasize that would-be exploiters or users of bodily products may very often be the more vulnerable parties to the transaction. After all, those who need donor tissue or organs very often stand to lose their lives if they are unsuccessful in their search for life-saving bodily products. As we saw in the last chapter, donors are almost always at less risk than recipients and usually the risks involved in live donation are, as we have seen, extremely small.

What is the moral objection to exploitation?

In the case we have been considering, that of the donation or sale of gametes or embryos or the leasing of a womb, the objection is not of course that the gametes are used as a means to another's ends (although they are so used) but rather the gametes' donor is the one exploited.

There are three issues. The first involves the question of whether it is morally objectionable to use others as a means to our own ends, the second asks whether paying them for being such a means constitutes exploitation, and the third is of course whether in all the circumstances, exploitation is such a bad thing.

First it is, I hope, obvious that there is nothing of itself objectionable in using others as a means to our ends. We do this all the time quite ethically and legitimately. The blood transfusion service is a good example. All blood donors are used by recipients as means to their ends. The ethic that requires us not to use people as means is derived from the Kantian principle that we should treat people not merely as means but always as ends in themselves. I am not sure precisely what this means but I suppose it is that we should not in our relations with others deny their personhood. We should not forget when treating them as means that they are also, and primarily, persons who matter morally. This ethic is not necessarily violated by the practice of blood donation nor is it by the donation of other body products.

It is not necessarily wrong to use others

We treat people as ends in themselves, as persons, when we invite them to adopt their capacity to help us, their contribution to ends of ours, as one of their own ends. One way of doing this, one way of trying to ensure that we do not exploit others, is to ask for their consent to what we propose and to make sure that they have a real option to refuse.

It is thus the vital issue of autonomy which bears importantly on one conception of the concept of exploitation. For it is the autonomy of a choice which pre-empts the claim that the choice was a use of one person by another which wronged the person so used. We can say that the Kantian demand that people be treated as persons, as morally important individuals with their own decisions to make and lives to lead, is not violated when they figure in our projects, if their doing so is also *a project of theirs*.

> It is not wrong to use others as a means to our own ends when they have autonomously adopted their part in our projects as one of their own projects and have not been coerced in some way into becoming instruments of ours.

In so far as some conceptions of exploitation involve nothing more than using others illegitimately as a means to our ends—the sense in which lovers may be said to exploit one another for example—we can provide a definition of exploitation in this sense:

> Exploitation occurs when those exploited have not autonomously adopted their part in our projects as one of their own projects but have been coerced in some way into becoming instruments of ours.

This important dimension of the idea of exploitation implies that the exploiter is able to apply some coercive pressure that those whom she exploits are unable or ill equipped to resist. The idea that when financial interests are involved, then using someone as a means to your ends is always exploitative, seems implausible. All so-called 'service' occupations involve this, from being a doctor to being a butler.

Suggestions that someone has been exploited usually seem plausible only when it is clear that there is some independent reason for regarding the particular, supposedly exploitative, practice as morally questionable.

If we continue the example of service occupations, we do not usually think that to pay someone to be a nurse or a household cleaner is necessarily exploitative, although it may be if the level of remuneration is not sufficient. But it is not the *fact* of remuneration that constitutes any alleged exploitation in such cases, but the *level*.

In the case of prostitution, however, it is often claimed that paying for sex necessarily exploits the prostitute. But this is not because payment is involved, but because there is a hidden assumption that prostitution is not the sort of occupation that anyone would voluntarily choose, *whatever* the level of remuneration. This hidden assumption neatly combines wrongful use and violation of autonomy. If prostitution is regarded as exploitative, it is because there are supposedly independent reasons for regarding prostitution as morally questionable, and hence a wrongful use of one person by another,[12] or the violation of her autonomy.

Clearly the thought is not that it is exploitative because it involves payment for something that ought to be voluntarily donated!

Those who regard prostitution or surrogacy as exploitative *per se* must then rely on the fact that there is a hidden assumption that because no one would voluntarily choose to be a prostitute whatever the level of remuneration, the choice cannot be autonomous. There is, however, a fall back position available here. It is that prostitution constitutes a wrongful use of one person by another not because the choice to be a prostitute is necessarily non-autonomous but because prostitution is *eo ipso* wicked and therefore a wrongful use of one person by another. I shall say no more about alleged wickedness of this sort. For a moral judgement to be respectable it must have something to say about just why it is that a supposed wrongful action is wrongful. If it fails to meet this test, as we saw in Chapter 2, it will simply be a form of prejudice and not a moral judgement at all.

One other candidate for an independent reason for objecting to a practice and regarding it as exploitative is not its supposed morally questionable nature, but the fact that it might involve danger. But again this is implausible as an objection to many occupations that we pay people to perform. There are occupational risks attaching to many respectable nonexploitative occupations—police, fire, army, and health professionals for example.

The first question here must be not is it wrong to pay people to face danger on our behalf, but is it wrong to pay people to face *this* danger? Or, possibly, is it wrong to pay them *so little* to face this danger? The second question is: have they been coerced into facing this danger?

These considerations show Warnock's simple assumption that commercial surrogacy is exploitative to be implausible.

It will be clear that Goodin's principle of protecting the vulnerable will not help us here either. For the unemployed as a class may well be prime examples of the vulnerable. The question of whether or not unfair advantage of the unemployed is taken by offering them employment is not settled by our knowledge of their vulnerability. Nor is it by our knowledge that those who offer them employment have 'flagrantly' pressed their advantage home.[13]

It is an odd fact that those who regard so-called 'altruistic' donations, giving away something or some service, as non-exploitative, might well regard being paid too little for it as a prime candidate for exploitation. Is it a paradox that being paid nothing for a service might not be exploitative whereas being paid something in the same circumstances might be?

I have suggested that consent pre-empts the claim that someone has been wronged by another's use of them. This is strictly true but it may not of course exhaust the wrongfulness of the transaction. We have noted that even such a consensual transaction may remain wrongful in a sense if there is some independent reason to regard the transaction as evil. Some of the possible evils which may remain have to do with justice. If in a transaction between A and B, B autonomously consents to be used by A, A may none the less gain unjustly and B may also have her interests set back by A's unjust gain. As Feinberg has noted: 'The evils that are produced, both by A's unjust gain and B's set-back interest, are non-grievance evils'.[14] By this Feinberg means that B has no justified complaint against A: B has not been wronged by A although he has been harmed.[15]

If B has no grievance and yet may still have been exploited does this demonstrate that we should object to the transaction? The answer must clearly be 'not necessarily'. We need to know whether on balance the benefits A derives should either on moral or public policy grounds be prevented or whether B should on similar grounds be protected against the voluntary damage she does to her own interests. We need to know whether or not, on balance, such harms as B suffers should be prevented in the light of, in particular, other goods that may come of the transaction.

Which brings us to the second element in our understanding of exploitation, or rather the second conception of exploitation. The first conception had nothing necessarily to do with commercial considerations and consisted simply, though not unproblematically, in the wrongful use of one person by another. We identified the wrongfulness of such a use as one in which the allegedly exploited individual had not autonomously adopted the supposed exploiter's ends as their own also.

In this sense there seems to be no reason to suppose, for example, that women who are paid, in coin or in kind, for donating eggs for therapy or research, have necessarily been exploited, nor that it is wrongful to pay them. The crucial question must be first: have they autonomously adopted this particular project as one of their own? If they have,[16] the only remaining question is whether or not the level of remuneration is such as to avoid the charge of exploitation. This is the second conception of exploitation to which we must now turn.

2. Exploitation and disparity of value

The second conception of exploitation necessarily involves some conception of a fair or a proper rate for the job or some view of the value of the product sold. A useful framework for considering this sort of exploitation has been given by Hillel Steiner. Steiner has suggested that the core characteristic of the concept of exploitation is that it involves a mutually self-interested, consensual exchange in which what one party transfers is—but need not have been—of greater value than what is received in return.[17] Steiner's account seems to me to be consistent with a Marxist theory of exploitation[18] but also usefully vague over just what it is that makes for disparity in value. It allows for any defensible account of the disparity in value involved in a particular transaction to be given.

Where there is an apparent disparity in value in an exchange of goods or services two obvious questions are: does the disparity constitute exploitation and if so why? And secondly, if it does constitute exploitation, what does this tell us about the moral permissibility of the exchange?

Disparity in value between one side of a transaction and the other begins to look like exploitation when there is some background injustice or violation of rights, or the transaction is wrongful in some other way. I shall not attempt an exhaustive account of just what makes disparity of value into exploitation.[19]

In other words, this second conception of exploitation as 'disparity of value' is dependent on the first conception as 'wrongful use'.

> Although there can be exploitative wrongful use without disparity in value, there cannot be exploitative disparity in value without wrongful use.

For our present purposes the question is whether, when and if such wrongful disparity of value can be shown to exist in the sale of bodily products, this constitutes a moral objection to the transaction? To answer this question we need to know more than that there is a prima-facie case of exploitation, we need to know whether it is wrong to permit such exploitation in all the circumstances of the case. In the light of these reflections on the concept of exploitation, if we wish to know whether a particular transaction is exploitative, we must ask the following questions:

1. Are there independent reasons for regarding the transaction as morally questionable? And are these reasons sound?
2. Are there features of the transaction that vitiate the autonomy of either party?
3. Is the transaction unacceptably dangerous given the benefits and the levels of remuneration?
4. Does the transaction involve a mutually self-interested, consensual exchange in which what one party transfers is—but need not have been—of greater value than what is received in return? And does this exchange involve injustice, a violation of rights, or some other wrongful feature?

The answers to these questions will indicate whether or not the transaction is exploitative. It is still of course a further and separate question as to whether or not such exploitation is on the whole impermissible in all the circumstances of the case.

II. Markets in Bodily Products

Back to gametes and fetuses

In the case of gamete donation, if it is acceptable to make voluntary donations of such things, then the addition of a financial interest does not necessarily add anything to the *morality* of the practice. Of course we must still ask whether the vendor has been exploited, and whether such exploitation is necessarily impermissible. But Warnock and Goodin are clearly in error if they think that financial interests either necessarily constitute exploitation or that exploitation is necessarily impermissible.

But have we forgotten something? Let's turn to the case of the use of fetal material. Here the ground has shifted somewhat, because whereas it seems plausible to think of gametes as part of the person's bodily products, the same does not seem unproblematically true of the fetus. At first glance it might seem that the fact that the mother's consent, indeed request, is necessary for an abortion implies that she somehow has also the right to determine precisely what happens to the fetus thereafter.

The role of consent

There are two aspects to the role of the mother's consent to what happens to her embryo or fetus. If it is planned and hoped that the embryo will be born and eventually become a person, then the role of the mother is that of the most obvious guardian of the best interests of the person the embryo will become. And her entitlement to consent to what happens to the embryo or fetus is contingent upon what she consents to being plausibly in that individual's best interests.

Where, however, the embryo will not be permitted to grow into a person (or in the case of an aborted fetus), the mother's consent seems only relevant in so far as it relates to what happens to her body. In the case of abortion for example, if the mother is entitled to an abortion, her entitlement is to eject the fetus from her body, not to pursue it thereafter. So that her entitlement is not *that it will die*, only *that it will be ejected from her body*. Here the ubiquitous Sisters of the Embryo whom we first met in Chapter 2, may have a role to play. If they wish to claim a viable aborted fetus it would seem doubtful that the mother had the right to pursue the fetus 'with an axe', so to speak, rather than let the Sisters of the Embryo adopt it.

Here the equivalents of the Sisters of the Embryo, are not those who might rescue the fetus, but those who might use it for therapeutic purposes. Here again, it seems ethically highly dubious to think that the mother might have a right to thwart the life chances of potential beneficiaries by declining to permit the use of her aborted fetus. She gains nothing of value commensurate with what others stand to lose by her actions. They stand to lose their lives or their chances of being restored to health.

Of course, though she would be morally wrong to refuse consent to the use of her fetus, and this, as I have suggested, would constitute grounds for denying her a right to refuse, by making the fetus, like cadavers, automatically available for therapeutic use, it does not follow that she should not be entitled to sell the fetus, or rather to be compensated for its being requisitioned by society.

Which brings us to a vital question.

Who has first claim upon the embryo or aborted fetus?

Suppose a woman who has an abortion wishes the aborted fetal material to be used to treat her father who suffers from Parkinson's disease. Is she entitled to claim the material from her fetus for her father, can she insist that he have first claim upon that material, rather than, say, the material passing into the public domain and all candidates for its therapeutic use competing equally for it?

Here it seems to me that this is simply an allocation issue. The main requirements of a principle of allocation are justice and economy. Someone should have the benefit of the therapeutic possibilities of the aborted fetus. It does not matter who, so long as the process of selection of the beneficiary is not unjust. It does not seem unjust that the woman's father should benefit from his daughter's aborted child rather than a stranger, and allocation to the father will certainly be straightforward and economical.[20] Of course it will not necessarily and always be the most economical use or the most morally pressing, and where this is so, these reasons for allocation elsewhere will have their due weight.

What of the abortion of a fetus who turns out to be viable? The mother wants it to be used for therapeutic purposes and the Sisters of the Embryo want to adopt it. Here I think the arguments of Chapters 2 and 3 indicate that the therapeutic use should be preferred. For the life and interests of an actual person are more important than those of a potential person, a pre-person. If abortion is justified to protect the life and/or the interests of an actual person, then so is the therapeutic use of an aborted fetus, viable or not. There is of course a sense in which the adoption of the fetus by the Sisters of the Embryo is therapeutic *for them*, but we will suppose not to the same degree as for the beneficiary of the fetal tissue.

But what if after the abortion the mother changes her mind and instead of wishing to use the viable fetus to save her father she wishes to keep the child? Her father on the other hand wishes to use it for therapeutic purposes. Are not the arguments the same? If the arguments for preferring the therapeutic use were sound, are they not still sound in these circumstances? I think they are, but the difference in the two cases turns on the cruelty of taking from a mother the child she wishes to keep. This seems an independent and enduring cruelty, the prevention of which would justify the sacrifice of the father's interests.[21] I place no weight upon the fact that the mother has changed her mind.

Finally, suppose the woman has become pregnant specifically to obtain fetal material for therapeutic purposes. Here the arguments seem entirely parallel with those we have just reviewed. A woman in this situation, it will be remembered, has no right to kill the fetus, merely to eject it from her body. If, once ejected, it is then of use or value to others, whether they be the Sisters of the Embryo or a transplant beneficiary, they should have it.

But might they be required to pay for it or to compensate the mother for the added distress or whatever that its ultimate destination may cause her?

Is the first claim a financial claim?

There seems to me no clear answer to this question. It is surely a matter of public policy not of morality, although of course the choice between rival issues of policy is often a moral choice. What matters morally is that the fetus should not go to waste at substantial cost to others. If payment is inimical to this objective then this provides a public policy argument against payment. If on the other hand payment acts as a necessary inducement or palliative to this end, then it seems to me to be, on the face of it, defensible.

Is it wrong to exploit the poor?

There is still of course a possible problem of exploitation here. The poor, or rather poor women, might be coerced into becoming pregnant in order to make a living by providing fetal material for transplant. There seems to be a clear moral imperative to prevent this. But is there?

Exploitation of the poor as donors of bodily parts would of course be undesirable. Whether it would be worse than many other things the poor are forced by their circumstances to endure I do not know. We are in danger here of wrongly assuming that if we prevent the poor from being exploited, this is the same as helping or caring for the poor. Or equally, that if we block their exploitation this discharges an obligation to the poor.

If we strongly object to the exploitation of the poor, as I do, then the therapy of choice is to remove the desperate nature of their poverty.[22] No one at all should be driven by poverty to do things that neither they nor society want them to do. However, in a world which is not taking independent steps to remove the poverty motive for selling bodily products or indeed anything else, it is not clear to me that we help the poor, or for that matter protect them, or show concern and respect for them, by preventing them from ameliorating their condition by selling some of the few things they have to sell.

A cartel in bodily products

Preventing the poor from selling body products might seem close to operating a cartel in bodily products from which the poor are excluded. One can imagine a reasonably well-off person, not exploitable and therefore 'free' to sell his kidney, retiring early on the proceeds of sale and living out his days in comfort, while the poor individual, debarred from selling bodily products, must work until he drops, or watch his family starve.

Who is my neighbour?

Nor does it seem obvious that we are doing right by our other neighbours—those who might benefit from the transplant material. They too are vulnerable, they too are entitled to our concern, respect, and protection. One way of according this to those who need transplants is to try to make sure that donor organs and other tissue are available to them either free, or, at a reasonable, as opposed to an extortionate, price for the body part in question.

We will be returning to this point, but we must now look at the special case of commercial utilization of children's bodily products.

The commercial 'exploitation' of children

At first sight the idea of the commercial exploitation of children's bodily products seems absolutely abhorrent. I hope it is clear, but I should perhaps emphasize, that children are a separate category here only and in so far as they are unable to give a valid consent to what happens to them. The moral distinction here is not that between adults and children but between autonomous and non-autonomous individuals.[23] I will, however, continue to refer to adults and children as a convenient but arbitrary way of signalling this distinction.

The commercial exploitation of children seems, as I said, abhorrent. But perhaps this is because we have described it in the most hostile and unsympathetic way. Now this is often a very good thing to do, for there is always something suspect about re-describing events until we find a palatable description. It may well be that we should not do anything *unless* we can face doing it under its most unpalatable description. This principle, if it is one, might prove a useful governing mechanism for morality. What is important is to be clear about just what would be involved.

The issue we need to discuss is whether, if children are to be used as donors of bodily parts, it would be wrong to pay them for these parts. If we were right to conclude that it would not be *unethical* for a parent to volunteer a kidney from one of her children to save the life of another child, would it be wrong for the donor child to be paid? We might well ask, could it be right for the donor child *not* to be paid?

Against a background in which commerce in bodily parts were permitted, it might well be judged that a parent who did not protect her child's financial interests would be manifestly negligent. The child would emerge as an adult to find that she had given away something for which others were receiving significant sums and might well wonder, since she had not been consulted about the donation, why her financial position had not at least been safeguarded?

Here again the main problem would be the effect which acceptance of the legitimacy of commercial dealings in children's organs would have on society. The dangers of exploitation of the poor would be great, perhaps too great to contemplate, although we should not too readily forget the claims of those whose lives might be saved by live donations. It may be that mechanisms could be found for avoiding the most likely dangers of exploitation or of mitigating its most harmful effects.

The crucial issue seems to be the permissibility of the practice in its non-commercial form. If the risk of being a donor is not too great a burden to place on a child, if it does not constitute an abuse, if it is not wrongful, then it seems doubtful that payment could make it so.

I do not want to champion or advocate commercial use of bodily products because my own instincts are all against it. I would prefer an entirely altruistic and consensual scheme. However, it also seems to me that we are too ready to emphasize the dangers, and too chary of considering the possible benefits of such schemes.

Protections for children

A number of fairly obvious protections in the case of the sale of bodily products of children and other individuals incapable of giving autonomous and valid consent might make such a scheme less obviously morally beyond the pale.

The following might be the beginnings of a set of morally respectable protections:

1. No consent to the sale or donation of children's bodily products will be valid where such bodily products are available from adults.[24]
2. No adult consent to the sale or donation of children's bodily products will be valid if the child is herself capable of giving or withholding a valid consent.
3. No adults shall be permitted to consent to the sale or donation of children's bodily products if they themselves are suitable vendors or donors and could supply the part.
4. In no circumstances will the sale of children's bodily products be permitted at prices less than those charged for adult products.
5. Trust funds would be established so that parents would not be able to benefit from selling their children's organs and the money would only go to children on majority.

Commerce in bodily products generally

If we compare the issues we have been considering with the situation of blood transfusions it might help to get things in perspective. In the United Kingdom we have a largely voluntary non-commercial blood donation scheme. In the United States there is a commercial scheme. I think that the United Kingdom scheme is altogether preferable. But it is not clear to me that the United States is acting *unethically* in operating a commercial scheme.

Of course there are disadvantages to a commercial scheme. People are tempted to sell their blood when their blood may be hazardous to others. Drug addicts, AIDS victims, and so on may sell blood when they would have no motive to give it away, and screening may not always pick up dangerous samples. However, these are considerations which make one scheme preferable to another. They do not show that it would be morally wrong to run a less than ideal scheme, particularly when the better scheme may not be viable in a particular society. Reasons for preferring one scheme to another are not necessarily moral reasons for preferring it.

Voluntary, commercial or conscription schemes

We could in principle have:

(1) voluntary donation;
(2) commercial market;
(3) conscription.

If scheme (1), the voluntary scheme proved insufficient then the optimal scheme would either be (2) or (3). Reasons for choosing between a commercial and a conscriptive scheme would be partially in terms of cost, but principally they would be moral reasons. We would use moral principles to determine whether it was better, morally speaking, to operate a commercial scheme or a compulsory one. Principles like individual liberty and autonomy might tug in favour of a commercial scheme, whereas principles like equality and risk sharing might pull in favour of conscription.

For present purposes I will assume that conscription is ruled out, on the grounds that a consensual scheme with inequalities is preferable to a compulsory scheme without, although there are sometimes substantial merits to conscription and I have discussed (and even defended) these elsewhere.[25]

Markets in adult bodily products

We have seen that voluntary live donations by adults are not ethically problematic. What difference would commerce make? Again the main danger is exploitation. If exploitation can be ruled out then again the choice between commercial and non-commercial schemes would be a matter of public policy. If, however, the shortfall in donor bodily products could not be made up by a voluntary scheme, this might well provide arguments for financial incentives.

We accept (perhaps we should not) substantial occupational risks for many people with substantially less moral justification than would be available to those selling organs to save the lives of others. It is not clear that the risks of tissue or organ sale are substantially greater than occupational hazards of other sorts, or that they are more vitiated by the financial interests involved than are the sale of services.

Safeguards

Of course if the sale of donor organs is to be permitted, it would need to be carefully regulated and questions about the level of remuneration and safeguards against wrongful exploitation must be carefully considered. In the recent United Kingdom case in which Dr Raymond Crockett was removed from the medical register for taking part in the sale of kidneys, allegations were made that the donors had not consented and indeed did not even know that their kidneys were to be removed. One Turkish donor, Mr Ahmet Koc, was reported as thinking 'he was undergoing a medical examination for [a] job but in fact he was operated on and a kidney removed'.[26] However, later the Turkish courts chose not to believe this story and he was given a suspended sentence in Turkey for trafficking in organs when 'it was found he had advertised his kidney in a newspaper'.[27]

It is difficult to know whether those who sell their organs would be wrongfully exploited or not. Leaving aside issues of autonomy and consent, the idea of exploitation seems here to imply some conception of a fair price. In so far as we do not know what a fair price for live donor organs would be, we have no reason to suppose that a particular price is unfair. For the allegation that a particular rate is exploitative presupposes either a fair price or at least a 'market rate'.

There is of course a certain problem for those who object to such sales of organs on the grounds of exploitation of the vendor and who use some conception of a market price. For they seem to be involved in the following problem: to deem a certain price exploitative (presumably of the donor—although an unduly high price exploits purchasers) and consequently to restrict sales prevents the formation of a 'market-place' and hence the establishing of the very 'market price' presupposed by judgements that a particular price is too low and hence exploitative!

Here again, even if we could be confident that in a particular case there was exploitation of the vendor this would not of itself show the transaction to be morally objectionable. We would need to know much more. If the decision to sell was not autonomous, then this would surely and clearly constitute a moral objection to the practice even at the cost of the life of the victim. But if the exploitation was merely a question of underpayment I doubt whether this would be sufficient grounds to ban the practice as a whole. Financial adjustments and compensation could always be arranged.

Here I have been concerned with the question of whether or not, in principle, the sale of human tissue and organs is ethically questionable. The practical implementation of a commercial scheme would of course require careful planning and detailed safeguards.

Because of the immense difficulties in the way of doing this successfully I would hope that sufficient donor organs could be obtained non-commercially. However, where this is unlikely to prove possible we should, it seems to me, consider alternatives.

Free market or regulation?

Whether there should be a free market with prices being set by the market or whether the price of body products should be nationally or internationally agreed is a further question, and one I raise only to ignore. We can imagine a price for organs at which there were no donors who would not have been voluntary donors, and we can imagine prices at which substantially greater numbers of donors would come forward. On the other hand of course there are clearly prices which few potential purchasers could afford and prices which most could afford. There is a clear sense in which while would-be purchasers are necessarily vulnerable, would-be donors are only contingently so.

Since it is desirable that those whose lives are at risk should be saved but no more lives be put at risk than is necessary to maximize lives saved, we should adopt the scheme which would have this effect. Which this is I have no idea.

Who is more vulnerable?

It is important to be clear about just what is and is not being asserted here and so I must re-emphasize a point made briefly already.

One question we should press here is, who is more vulnerable, who is more in need of our protection? If we ask this question we might see the ethics of commercial transplantation in a different light. The dying people who need transplants are also entitled to our concern, respect, and protection, they do not wish to die. Those who would choose to sell organs are volunteers to a small but significant risk. Is it morally preferable to subject one group of citizens to certain death rather than offer incentives (temptations if you like) to another group to run risks? Isn't it rather better to protect the most vulnerable by permitting another group to choose whether or not to run a risk in the hope of both benefiting their fellow human beings and benefiting themselves financially?

Is a society which allows 20,000 people a year to die for want of donor organs a better society than one which saves those lives by allowing other citizens to run small risks to prevent such deaths and which pays the citizens for so doing?[28]

In a sense we already live in such a society. Any society which maintains rescue services and pays rescue personnel to run risks *is such a society*. When a society maintains fire services, police and ambulance services, military 'defence' forces, and even health professionals it accepts that it will call upon such personnel to run risks, including risks of death in the public interest and it accepts that such people should be paid for so doing whether or not they are volunteers.

The ultimate sacrifice

Just as it seems obvious that there is a price at which extra donors of non-vital organs like kidneys or non-vital tissue like bone marrow will come forward, so, doubtless there is a price at which donors of vital organs like the heart will offer their organs for sale. Should this be permitted? If I may give my heart away why shouldn't I sell it? Again people driven to such an expedient must be in dire straits and we would hope for alternative remedies.

But if it isn't wrong for me, for example, to give myself up to certain death to save others, then if only payment of a ransom (or its moral equivalent) will save those others and my heart can purchase the ransom why shouldn't I do it? Of course, under no circumstances would the donation or sale of vital organs or other bodily parts be permissible from persons incapable of themselves deciding about and consenting to such a procedure.

A possibility which appears to make this last claim over hasty is the case of a terminally ill individual in such dreadful pain that euthanasia seems highly desirable and who has poor children who can benefit from the proceeds of organ sales. But here the feature of the situation which legitimates the donation is the desirability of euthanasia. If euthanasia is decided upon, then it is this that legitimates the taking of organs and the individual can certainly be kept alive long enough for organs to be taken. This case would then be like many 'normal' cases of cadaver transplants.

Conclusion

I have tried in this chapter to survey a range of issues concerning the ethics of using the human individual as a commercial resource centre. In doing so I have inevitably covered too much ground too rapidly for a full discussion. This has been unavoidable because of the vast complexity of the issues touched upon. Although many of the concerns of this chapter have been peripheral to the main argument which, readers will be forgiven for having forgotten, is about the ethics of human biotechnology, they hover on the sidelines and at least some consideration of them was required. We must, however, now return to the central questions concerning the ethics of specifically genetic manipulation of human beings.

7

Wonderwoman and Superman

'Mens sana in corpore sano'—'a healthy mind in a healthy body'—sounds like a school motto. It may for all I know be the motto of a thousand schools or none. Few would I suspect quarrel with it as a statement of an important part of the aims of education, unless of course they objected to their students being inculcated with a cliché. But suppose a school were to set out deliberately to improve the mental and physical capacities of its students, suppose its stated aims were to ensure that the pupils left the school not only more intelligent and more physically fit than when they arrived, but more intelligent and more physically fit than they would be at any other school. Suppose that a group of educationalists, outstanding ones of course, far more brilliant than any we know of to date, had actually worked out a method of achieving this? What should our reaction be?

Well of course our reaction would be one of amazement, it would certainly be an unprecedented event—a breakthrough in education. But should we be pleased? Should we welcome such a breakthrough? We might of course be sceptical, we might doubt such extravagant claims, but if they could be sustained would we want our children to go to such a school? And if the school our own children attended was not run according to the new educational methods, would we want these to be adopted as soon as possible?

We ought to want this. It is, after all, what education is supposed to be for. Indeed, if the claims were not expressed hyperbolically and competitively, it is what, if we knew little about education, we might well imagine was actually going on in schools or at least was what the teachers were supposedly trying to bring about. Of course we might have some reservations. We might want to be assured that others of the things we want from education would not be sacrificed in the cause of intelligence and bodily health. We would want this school and these educational methods to transmit the culture (or the multiculture as we must now think of it), and we would want our children prepared for the real world. But this said, if the gains in intelligence and health were significant and palpable we might well be willing to postpone initiation into some elements of our multicultural heritage or forgo some extra periods of personal and social education for the sake of these compensating gains in intelligence and health.

And, if we had reason to believe that these other educational objectives were all the more likely to be successfully achieved the higher the intelligence of the students, then it would not be a case of forgoing this dimension of education but merely of postponing it. We could then continue education from a higher base so to speak. And if the improvements in health indicated longer expectation of life, we could take comfort in the thought that we could well afford to spend longer on the educational process than has been customary.[1]

Now we can again entertain conjecture of a different sort of breakthrough with the same or comparable consequences and suppose the new biotechnological procedures could engineer into the human embryo characteristics which would make highly probable the expression of adult phenotypes like build, height, and even intelligence and could also reduce susceptibility to disease. The combination of build, height, and reduced susceptibility to disease would make highly probable the production of a healthier, fitter individual, and intelligence might also be susceptible to engineered enhancement. Of course, it would be very difficult and hence unlikely that we would be able to influence intelligence significantly. Intelligence is a very complex matter, multidimensional and multifactorial, and to influence even one of its dimensions would almost certainly involve the manipulation of too many genes to be viable in the foreseeable future. However, the principles that the possibility of such modifications raises are important and it is as well to try to be clear about them now, long in advance of any possible manipulations.

This might happen in a number of ways. Genetic probes could be used to screen for likely adult phenotypes (build, height, intelligence, etc.) and only promising embryos would be implanted or permitted to go to term. Others of the possibilities we have already examined in earlier chapters would provide opportunities here. For example, it might also be possible to carry out genetic screening at the gametes stage and only fertilize promising eggs with promising sperm. More technologically dramatic would be the possibility, still remote, of engineering additional genes into the human genome. These genes might for example be coded for antibodies to all the major infections and so radically improve the health expectations of the individual. They might also be related to adult phenotypes such as intelligence. Let's not worry too much about the mechanism by which these improvements might be produced, we'll leave that to the bioengineers who work in the selfless hope of a Nobel prize. Our question is this: if the goal of enhanced intelligence and better health is something that we might strive to produce through education, including of course the more general health education of the community, why should we not produce these goals through genetic engineering?

If we could engineer enhanced intelligence and health into the embryo should we not do so? If these are legitimate aims of education could they be illegitimate as the aims of medical, as opposed to educational, science? This is, perhaps surprisingly, a very complicated question and one of immense importance for a number of reasons. Biotechnology is not of course ever going to be a substitute for education. Even if, *per impossibile* perhaps, we can engineer in higher intelligence or the probability of higher intelligence, the beneficiaries will still require education, though we may expect the educational process to be more rewarding all round. But of course, educationalists have a large interest in whether or not such technological enhancement of abilities is to be implemented. And as citizens we all need to know whether or not we ought to welcome such prospects, and if we are inclined to welcome them, we need to know whether we are right to do so.

Less prosaic is the underlying question of whether we might be, and are entitled to, change the nature of human beings. And of course considering this question reminds us of the interesting and no less important issue, essentially philosophical, of just what it is to be human. A question which goes, of course, to the heart of education. Before we can address these issues we must first try to identify the various important complexities of the question of the legitimacy of these procedures.

Changes of degree and changes of kind

We have seen that genetic changes may operate on the germ line or on the somatic line, that is they may be one-off changes to individuals or they may enter the genome of the individual and be transmitted to any offspring of that individual. There may be different advantages to both methods. There may for example be the necessity to repair, if we can, damage to a particular individual, and the repair may have consequences it would be undesirable to pass on to future generations. On the other hand, in the case of an inherited defect, rather than have to carry out repairs in each future generation it is more effective in every way to eradicate the defect once and for all by operating on the germ line. However, again as we have noted, operating on the germ line makes a permanent alteration to the genetic make-up of the individual. This is of course a momentous thing to do for it carries with it a number of dangers. The first is that any undesirable side-effects of the genetic manipulation are built in, and successive generations are condemned to experience these side-effects.

Moreover, while reducing our susceptibility to disease and enhancing intelligence may obviously be humane aims, their successful realization may have profound consequences for our understanding of just what it is to be human. Species of course do evolve, the question is whether they evolve as a species or whether one species will in fact evolve into a new and different species.

The Shylock syndrome

Shakespeare's *The Merchant of Venice* is at times profoundly ambiguous on the issue of racism. But when Shakespeare gives Shylock his most sympathetic speech, he defines humanity precisely in terms of a community of traits, and the most significant traits are human frailties.

> Hath not a Jew eyes? Hath not a Jew hands, organs, dimensions, senses, affections, passions? fed with the same food, hurt with the same weapons, subject to the same diseases, healed by the same means, warmed and cooled by the same winter and summer as a Christian is? If you prick us do we not bleed? if you tickle us do we not laugh? if you poison us do we not die? . . .[2]

The profound question here is: are there traits which are constitutive of simply being human as we understand the term? Clearly there are, though I will not be so imprudent as to attempt to specify just which they are. Three questions immediately arise. They are: what are these traits? How profoundly might they be changed and we yet remain human? And finally, would we, or ought we to, want to remain human if the price of the loss of our humanity was dramatically enhanced health and intelligence?

Perhaps we have conceded too much already? When smallpox was eradicated we did not think that we had become less human for having succeeded in wiping out such a dreadful disease. So why might we be more worried about reducing our susceptibility to other diseases? Perhaps the answer is simply that in eradicating smallpox we have not changed our susceptibility to it. We have not changed ourselves, rather we have changed the world so that it does not include smallpox (except of course in medical laboratories). Those who take comfort from this way of putting the matter make it begin to look as though we might want human beings to remain vulnerable to the disease so we would have the advantage of succumbing once again should smallpox recur.

But this is not quite right either. For smallpox has been eradicated in large part because vaccination has been successful. And vaccination is a somatic modification in the sense that it modifies only the particular individual vaccinated, so it was the removal of our susceptibility to the disease that was instrumental here. It looks as though we have again come back to the difference between a permanent removal of that susceptibility and a temporary one. Again we seem to be worried by the idea of permanent changes to human beings, perhaps because we fear this may change human nature?

Would a Shylock, of his nature less susceptible to bleeding when pricked, be less human on that account? It would be dangerous to answer 'yes', for there are many diseases or conditions which affect some racial groups but not others—sickle cell anaemia for example. So that to conclude that the advantage of immunity to a danger which confronts others made one less human would already involve profound divisions within the human community over and above those arising from sheer bigotry. Of course, this does not show that systematic modifications to the human genome, of a fairly comprehensive sort, might not involve changes in kind as well as changes in degree. If we managed to create a strain of so-called 'superbeings' or 'superhumans' who were comprehensively less vulnerable and perhaps also more healthy, intelligent, and longlived, this might so disturb our familiar and cherished categories as to make us regard such beings as inhuman. But so what?

There now remain really only two questions, but they are of the first importance and are very hard to resolve. The first may be modified from Shylock's own next question. It is: might we wrong people by changing their genetic susceptibilities and might they not then rightly seek revenge? The second is more subtle. It is implausible to suppose that we might wrong those whom we change, for the changes will only be even prima facie justifiable if we might thereby advantage those whose human frailties we minimize. The question we must ask is: might we wrong humanity at large by either creating a new subspecies of superbeings, or indeed and eventually, changing humanity entirely for we would believe to be the better?

These questions are simply put but are complicated to resolve. There are very many different issues at stake here and each constitutes a possible objection to the genetic modifications we envisage. In this chapter we will concentrate mainly on the intermingling of human and animal genes and other parts; in subsequent chapters we will look at the ethics of creating an entirely new breed of persons.

I. Hybrids

More radical and more obviously threatening to our humanity are changes that involve the introduction of material from other organisms—so called 'transgenic' modifications which in the contemporary science fiction of science documentary programmes means bizarre hybrids. There seems to be an instinctive hostility to such practices and we must ask whether this hostility is well founded.

Much has been made of the creation of transgenic animals and most people have had the chance to see the results on television, with strange creatures produced from crossing sheep and goats with the resultant wailing and gnashing of teeth, usually from outraged viewers. A transgenic animal or plant is simply one genetically modified by the introduction of DNA sequences from another organism by methods other than conventional breeding. In a phrase, transgenic creatures contain non-parental DNA. Fuel has been added to the flames of controversy by a recent ruling in the courts of the United States to the effect that a transgenic animal can be patented as a 'new life form'.[3] This is a potent phrase conjuring up pictures of biologists supplanting God's monopoly in the production of new life forms and leading to expectations of God's instructing leading counsel to appear on her behalf and restore the priority of her own patents.

Leaving to one side the problem of blasphemy and the question of whether or not the Almighty's patents are still current, there remains the question of whether or not there is anything wrong with creating new life forms by transgenic modification. There seem to be two elements to this sort of objection to genetic modification. The first has to do with the idea of usurping God's prerogative and creating new life forms or perhaps disturbing the course of nature. The second element has to do with the horror and perhaps the taboo attached to crossing species boundaries.

Playing God

If it is supposed that we ought not to play God a number of assumptions must be made. The first is that God has a monopoly of the role; the second is that she is doing a good job (or a better one than we would do) and perhaps in consequence has a right to be left to get on with it; the third is perhaps that God's will is expressed in nature and that consequently the so-called natural order must not be disturbed. You don't have to be an atheist to see that the idea that we ought not to play God is a non-starter. Even believers must believe it can be right to disturb and redirect the course of nature otherwise the practice of medicine itself would be wicked. For people naturally fall ill and naturally have reparable defects; if the practice of medicine has a coherent aim it must be seen, if anything, as the comprehensive attempt to frustrate the course of nature. No one who believes it right to take an antibiotic or to vaccinate her children believes either that God is doing a great job unaided or that it is wrong to disturb the natural order. Moreover, the idea that human beings should not disturb what God has so carefully arranged presupposes that we and the disturbing things we do are not part of those arrangements.

Crossing species boundaries

There are two separate issues here. The first is the question of whether it is wrong to cross species boundaries by creating hybrids, mixtures of the genetic material of two or more species to produce genuine chimeras. The second issue is that of whether it is wrong to dilute the pure humanity of human beings by an admixture of material from other species. It is not easy to keep these two issues apart but the possible objections are of a different order so we shall try.

1. Chimeras

The original Chimera, as killed by Bellerephon, was a remarkable hybrid, having a lion's head capable of breathing fire, a goat's body, and a serpent's tail. It was in short a monster, and as such was of course female. The term is now used generally to cover hybrids where the result is clearly identifiable as an oddity, the sort of creature that used always to be called 'a monster'.[4] Now there are three sorts of objections to creating monsters. The first is they frighten the horses and perhaps also ravage the earth; the second is that they are disturbing, they make us uneasy; and the third that they themselves are likely to be miserable. The solutions to some of these supposed problems are of course easy: we can keep them away from horses and if they look like ravaging the earth we can prevent or encourage them as seems appropriate. As for making us uneasy, well, we have no sacred right to tranquillity of mind, although many people think that they do. This objection we will consider again when we look at the second dimension of the problem. The most compelling objection to our creating monsters is that they themselves will be unhappy.

There are two possible dimensions to this unhappiness. The first is that the nature of the hybridization (if there is such a word) will make the resulting creature uncomfortable, or even cause it pain simply to exist. This might happen if the sort of body that it ends up with is maladapted to existence in the world so that, for example, it is painful for the creature to walk, or eat, or whatever, simply in virtue of the configuration of limbs or other parts of the body that it finds itself blessed with. The second source of unhappiness for chimeras stems from their oddity. They will be unlikely to have other creatures to relate to, although in principle a new species could be manufactured to overcome this problem. More importantly, they are likely to be treated as monsters—at worst they will be rejected and at best be made the objects of intrusive curiosity and perhaps contempt.

Clearly there will be an important difference in people's likely reactions here depending on whether or not the hybrid is or is not part human and an important moral difference depending on whether or not the hybrid is or could become a person.

The arguments about the morality of creating purely animal hybrids are related to arguments about using animals for experimentation, drug testing, and the like. We should not cause pain to sentient creatures unless there are compelling reasons so to do: reasons, moreover, which reveal that something so morally important is at stake as to justify the wrong of inflicting suffering on sentient creatures.

2. Humans and animals

Would it be wrong to produce a human hybrid or indeed a transgenic human? Again we must be careful to be clear. Species boundaries between humans and animals may be crossed in a number of ways. Transgenic humans might be produced by introducing DNA from another creature into the human genome. This might result in a small modification leaving the resulting individual essentially 'human'. Secondly, tissue or organs might be used for transplants from animals to humans—this has already happened extensively with human beings. As far back as 1968 a team at the National Heart Hospital London tried to save a patient by linking his heart to that of a pig. Since then human patients have been given or been connected up to the hearts of both sheep and baboons. In 1984 the celebrated 'Baby Fae' died 20 days after being given the heart of a baboon.[5] These techniques have been used for a number of years now and the use of insulin derived from pigs for the treatment of diabetes has, for example, for many years been standard practice. More sensationally, humans and animals might be cross-bred, either by conventional animal husbandry techniques or by genetic engineering.

When, in 1988, a team at Dulwich Hospital announced they had discovered a way to prevent rejection by humans of animal tissue and organs which might pave the way for routine organ transplants between species there was an immediate outcry. The main objection was from animal defence groups, one of which was reported to have said 'the idea of breeding pigs as living organ banks owes more to sick horror fantasy than medical science'.[6]

The objection to using material from animals to save human lives or to ameliorate the condition of human beings on the grounds of the wrong thereby done to animals has in my view no force at all, although I shall not argue for this conclusion here and now. It derives from the moral difference between most humans and most animals. While we should never cause sentient creatures pain or distress if this can be avoided, it is not wrong to take the lives of creatures that cannot value existence (non-persons) and so to use, for example, animals to benefit humans who can value existence (persons). In any event, no persons who are not vegetarians can have this sort of objection to using animals for therapeutic purposes.[7] As the doctor who released the information about the Dulwich breakthrough said: 'I would be unhappy if the Animal Liberation people felt there was a difference between taking a pig's kidney for transplant purposes and using it to feed the cat.'[8]

The other sort of objection to the insertion of animal derived material within the human body has more to do with taboo. It is a sort of instinctive recoiling from the very idea of contamination with animal products. Like other taboos which employ the ideas of defilement and uncleanness it is either felt or not. It deserves about as much respect as objections to miscegenation. In the absence of an argument or of the ability to point to some specific harm that might be involved in crossing species boundaries, we should regard the objections *per se* to such practices as on a par with objections to interracial marriage and dismiss them as mere and gratuitous prejudice. We will, however, return to this issue and to ideas about contamination or to the objections to violating cherished categories a little later in the course of our present discussion.

3. Human chimeras

We can come now to the most sensational dimension of this problem, the possibility of creating human chimeras properly so called. It is difficult adequately to specify what is meant by 'properly so called' in this context. I would like to be able to stipulate that this simply means humans so transgenically modified or hybrid that either they no longer look completely human or that they would no longer be generally accepted as human beings. However, transgenic modifications of a quite radical sort might be achieved without any outward show. Moreover, as we know only too well, genetic defects or indeed accidents can produce very dramatic and 'inhuman' appearances in bona fide members of the human species. There is then a continuum between small transgenic modifications and what one might call full-blown human chimeras of radical appearance and perhaps also attributes. This fact can be important in arriving at a balanced view of what our response to the prospect of creating human chimeras should be. Perhaps we can better approach this problem by considering some possible sorts of cases. But first let's just put the problem in some sort of historical perspective.

Teratology, or the study of monsters, has always had a certain fascination. A recent medical text on congenital malformations devotes an entire chapter to the teratology of the past.[9] Writing specifically about the history of our response to human hybrids, Josef Warkany reminds us that there was in the past a widespread belief that members of different species might be able to breed with one another and that reactions to the possibility of humans choosing to breed with other species varied widely. Greek mythology is, as we have noted, full of examples of half-human creatures which were not necessarily repulsive. Warkany reminds us that in ancient India and Egypt a 'cross between different species was not considered repulsive, and if a woman produced a monstrous child who resembled a particular animal, the child was paid the same respect due that animal in the official religion. For instance at Hermopolis in Egypt, the mummy of a human anencephalus was found in a grave reserved for sacred animals.'[10]

Evans-Pritchard's celebrated study of the Nuer[11] indicates that related practices have persisted to this day.[12] The trouble, as far as the acceptance of hybrids and their parents goes, begins, as do so many problems, with the Christian gloss upon Mosaic law condemning sexual relations between humans and animals. Again, Warkany notes that the 'Danish anatomist Bartholin mentions in his writings that a girl who gave birth to a monster with a "cat's head" was burned alive in the public square of Copenhagen ... This happened in 1683 in a civilized country.'[13]

Let's consider the problem of hybrids and the possible moral constraints on their production by looking first at a perhaps extreme and extremely unlikely possibility. In the United Kingdom, viewers were shocked by a television drama featuring the straight cross between a human and a gorilla.[14] The hybrid, produced supposedly by old-fashioned artificial insemination by donor, was the product of government sponsored genetic engineering research. The calls received by television stations wanted to know whether such a thing was possible, whether it was actually happening, and why it wasn't being stopped. As it turned out in the programme, the resultant baby soon shed its extra hair and looked perfectly human but viewers were not comforted by its normal baby-like appearance. Would it be wrong to do such a thing? What objections might there be to creating a half-human hybrid?

A straightforward attempt to cross a human with a gorilla is a gross procedure. It would be very difficult to predict just what form the resultant individual might take and consequently difficult to predict what the advantage might be in doing such a thing. The reasons offered in the programme were conflicting and completely inadequate. The 'scientist' who carried out the procedure dreamed of creating a new race, with man's intelligence but without his murderous aggression. The government representative speculated about three possible uses: for space travel, for organ donation, and to provide hybrid battalions for the army with 'total loyalty'. The inadequacies of these reasons are interesting.

First, the dream of the new pacific race is a non-starter. Apart from practical problems which include the likelihood of hybrids being infertile, there is simply no possibility of such a race replacing human beings, so the problem of human aggression would not thereby be solved. All that could be achieved is the creating of a new species alongside ourselves, which would doubtless have its own problems, not least of which would be how the new species would protect itself from the designs of *homo sapiens*. Hybrids could I suppose be used for space travel. But if they were intelligent enough to count as persons (a very minimal intelligence indeed[15] judging by the human persons who everyone agrees do count), and they would have to be to be useful in this role, then again they could not without their consent be used either for space travel or as organ donors any more than you or I can be. For if they were to develop into self-conscious beings they would constitute a non-human species of persons, but as persons they would have the moral status of persons and would share with us as the only other known species of persons the rights, freedoms, protections, and obligations that persons possess in virtue of their personhood. Battalions of hybrids could indeed be created as battalions of humans so often are, but again there is the problem of the preferences of the hybrids themselves. The world has unfortunately never lacked for willing soldiers, and the idea of 'total loyalty' is just plucked from the air. And again, loyalty amongst armies has not proved a particular problem.

The interest in and difficulty of finding adequate justifications for such an experiment highlights the point that while morally neutral actions do not require justifications, morally consequential actions do. What makes the prospect of producing a human hybrid morally consequential? Who might thereby be wronged? The idea that it might be used for wicked purposes does not make the act of producing it wrong. When maniacal Nazis copulated with a view to producing 'soldiers for the Reich' (if they ever did), this did not make procreation wicked. There may be perfectly nice individuals walking the streets of Hamburg today who were engendered in just such a spirit. We could coherently permit the production of hybrids and simply block the wicked exploitation of them. But most people would think there was still something wrong with producing a gorilla/human regardless of the use to which she might be put.

The wrong of experimenting on persons

One possible candidate for the wrong done in a case like this is the idea that we ought not to use human beings as guinea pigs. Now of course there is the hint of a paradox here for a creature that is half human and half derived from another species is not, by definition, human. However, it is only the hint of a paradox because only a scoundrel would attempt to resolve a moral problem by stipulating it out of existence. What is meant by the spirit of the objection is that morally important creatures like humans, persons who are not humans, should not be experimented upon. This is not the place for an exhaustive account of the rights and wrongs of experimenting on persons. However, one or two points can usefully be made. It is usually thought wrong to experiment on people without their consent because this is a violation of their autonomy. Where consent cannot be obtained, as with embryos, young children, or those unconscious or insane, the wrongness of experimenting on persons turns on the wrongness of harming them. The supposition is that an experiment is an experiment. It may do good and it may not. If we should not harm people, we should not, for the same reasons, expose them to the risk of harm. Now this of course applies to experiments on pre-persons, individuals who are not yet persons but who will in all probability become persons.[16] The wrongness of harming pre-persons does not lie in the wrongness of harming potential persons but rather in the wrong of harming the actual persons that the pre-persons will become.

This does sound embarrassingly like the grossest sophistry, but it is not, or perhaps I should say that I hope it is not. The point is simply that it is not wrong to harm potential persons so long as the potential is never actualized. As we have seen,[17] frustrating potential, as in the case of abortion or selective termination, is not wrong. So that whereas pre-persons are of course potential persons their moral importance lies not in their potentiality but in the fact that the potentiality will be actualized. And that when it is, harms done at the pre-person stage will be harms done to the actual person she becomes. It is a form of delayed action wrongdoing.

So, the wrong of producing a hybrid might be the wrong of harming, or causing suffering to the hybrid, if the fact that it is a hybrid will have this consequence. If this is right, important consequences for non-science-fictional practice will flow.

It is wrong to produce disability

If the production of the hybrid we described a while ago, the half-gorilla half-human hybrid, is wrong, it is so because we ought not deliberately to produce a creature that will very probably suffer. We should not in short bring avoidable suffering into existence unless of course there is a very powerful moral reason for so doing.

Now the power of the moral reason required will vary relative to the moral importance of the creature that will consequently suffer. The wrong of causing suffering to persons is usually worse than the wrong of causing suffering to other creatures because the suffering of persons always has a psychological dimension. Persons suffer in anticipation of suffering, in memory of suffering, and because they are conscious of relative deprivation of freedom from suffering. So the wrong of causing suffering to persons is usually greater and a greater wrong is done when that sort of suffering is gratuitously produced. So that if hybridization will produce a monster, a creature that is likely to suffer either because the physical configuration is not conducive to a pain-free or otherwise unrestricted existence, or because it is unlikely to survive for long, then deliberately to or knowingly to create or risk creating such a creature is, if it comes to exist, to do it wrong.

But in these respects, possible hybrids do not differ from many existing human beings. If it would be wrong knowingly to create a hybrid that would be likely to suffer, then the same wrong is committed when a human being who will suffer is deliberately or knowingly produced. This is the case where human parents might wrong their children by bringing them into existence, which we considered earlier. To produce a hybrid human is no more nor less wrong than producing a 'thoroughbred' human who will experience comparable degrees of pain, suffering, or restricted existence.

We have concentrated on the probable pain, suffering, or restricted existence that might result from hybridization and compared this to comparable conditions found in human beings. One immense problem that would face a half-human hybrid is simply the fact that it would be thought of as at best an oddity and worst a monster, whether or not it could be considered handicapped or disabled in any other way. I am imagining the case of a perfectly healthy, mobile, pain-free hybrid whose only problem in life is how it will be perceived by others.

Here we seem to run up against incompatible views. On the one hand we feel that there is no respectable defence of those who reject others or make their lives more difficult simply because of the way they look or who reject them because their identity is ambiguous. Here ambiguity of species identity is related to say ambiguity of gender identity or race or ethnic group identity. If we think it wrong to discriminate against people or reject them or subject them to abuse because they are of mixed race or gay or hermaphrodite and so on, it can be no more morally respectable to visit these evils upon possible hybrids. On the other hand we know that to be disfigured or deformed is an immense disadvantage and we rightly sympathize with those who suffer in this way and wish to help them achieve the looks that they themselves would wish to have.

The ground here is very slippery. We must of course say that to be of mixed race or to be gay is not a deformity nor yet a disfigurement and those who react adversely to gays or people of mixed race are mere bigots. But while accident victims who have say, facial burns, are literally disfigured, their subjection to adverse reaction by others is no more morally respectable nor less reprehensible. The distinction, if there is one, between these two cases lies in how those who are the subjects of adverse reactions think of themselves. Usually, gays and those subjected to adverse reaction on grounds of race or racial mix are happy with themselves and only lack acceptance by others, so the remedy is to work on the others, to show their adverse reaction to be indefensible and shaming. Those who are disfigured or deformed on the other hand would very often wish to look and be different, they would wish to be without their disability (for want of a better word), even if they were completely accepted by others and in no way restricted by how they look. Of course this claim is itself suspect since in an atmosphere of complete acceptance disfigurement might have ceased to be disfigurement. Our problem is that if we were to produce a hybrid human in the absence of other such creatures, we could not know how the hybrid would himself feel. I submit, however, that we can make a very reliable guess that the first hybrids will wish to be normal human beings and will be subjected to prejudice in the community at large. At least they will if hybridization carries no compensating advantages. So that where we know that a particular individual will be born 'deformed' or 'disfigured', whether because of hybridization or not, the powerful motive that we have to avoid bringing gratuitous suffering into the world will surely show us that to do so would be wrong. The problem, of course, arises where the suffering would not be gratuitous in the sense that we might be able to produce compensating advantages through hybridization to counteract the supposed deformity. This dilemma we will consider when we come to look at the advantages which we might be able to engineer into new persons. But one further possible objection to hybrids remains to be considered.

Human hippopotamuses

It is difficult not to be influenced by the writing of Mary Douglas on these matters. In her influential book, *Purity & Danger*,[18] Douglas suggests that our ideas of pollution or defilement are connected to our sense of order and place, particularly perhaps our sense of the natural order. '[I]f uncleanness is matter out of place, we must approach it through order. Uncleanness or dirt is that which must not be included if a pattern is to be maintained.'[19] Thus: 'Shoes are not dirty in themselves, but it is dirty to place them on the dining table; food is not dirty in itself, but it is dirty to leave cooking utensils in the bedroom . . .'[20] Citing as her source Evans-Pritchard's work on the Nuer she notes:

> For example, when a monstrous birth occurs, the defining lines between humans and animals may be threatened. If a monstrous birth can be labelled an event of a peculiar kind the categories can be restored. So the Nuer treat

monstrous births as baby hippopotamuses, accidentally born to humans and, with this labelling, the appropriate action is clear. They gently lay them in the river where they belong.[21]

Commenting on Douglas's approach Jonathan Glover, in his pioneering work on the philosophy of genetic engineering,[22] makes three points. He notes that 'to the extent that our resistance to genetic mixing is caused by revulsion against anomalies it is tempting to dismiss it'. This temptation has two dimensions. It is the temptation to dismiss taboo as mere prejudice, a temptation to which I think Glover is, like myself, ready to submit. But Glover enters the caveat that this taboo is probably 'something we could give up without loss'.[23] He continues, however, by noting that it 'may not be easy to give up our feelings of revulsion' and likening it to our own feelings of revulsion about eating particular edible but to us unsavoury foods, he concludes that we 'know rather little about how deep this kind of reaction goes and should not assume that shedding taboos is without psychological cost'. Making his third point Glover continues:

If instead of there being a clear gap between monkeys and ourselves, genetic mixing resulted in many individuals varying imperceptibly along the continuum between the two species, this might undermine our present belief in the moral importance of the distinction. If it did, the effects might go either way. There might be a beneficial reform in our attitudes towards members of other species. Or there might be a weakening in the prohibitions that now protect weaker or less intelligent human beings from the treatment animals are subjected to. If the second possibility is a real danger, there is a strong case for resisting the blurring of this boundary.[24]

It is difficult to know quite what weight to give to considerations such as these. On the one hand we do not want to incur unnecessary costs including psychological costs. There is no point in trying to do something just for the sake of it or to see what will happen when there may be hidden dangers. On the other hand, when there is a strong possibility of achieving something important, we should not forgo that opportunity for the sake of a rather nebulous possibility of some psychological unease.

Glover's third point is even more difficult to assess. He posits a change to which there might be a benign and a malicious response and seems to suggest that we take a worst case scenario and, if there is any danger of malice ruling, we should regard a strong case for conservatism as having been made out. There is some resonance between these suggestions and the principle of argument lampooned by F. M. Cornford in his *Microcosmographia Academica*.[25] Cornford's 'Principle of the Dangerous Precedent' is:

> that you should not now do an admittedly right action for fear that you, or your equally timid successors, should not have the courage to do right in some future case, which *ex hypothesi*, is essentially different, but superficially resembles the present one. Every public action which is not customary, either is wrong, or, if it is right, is a dangerous precedent. It follows that nothing should ever be done for the first time.[26]

While this would be grossly unfair as a criticism of Glover's rational caution, it contains the germ of an important point. This is simply the reminder that the reasonableness of bowing to adverse reaction, which may either be prejudice or a simple gut reaction, whether or not prompted by taboo or the disturbing of basic categories, depends on whether or not the actions or practices objected to are right, or, at least not clearly morally wrong. For if, as Ronald Dworkin has persuasively argued, particular practices can be shown to be immoral, 'then the freedom to pursue them counts for less. We do not need so strong a justification, in terms of the social importance of the institutions being protected, if we are confident that no one has a moral right to do what we want to prohibit.'[27] Dworkin was talking about homosexuality and criticizing Lord Devlin's famously intemperate hostility to it, but this is also a significant reminder. If hostility to hybrids is like that to homosexuals or to particular races or racial mixing, if these things are right or at least not morally wrong, then the good person will try to eradicate her prejudice, and that of others, rather than the practice in question. And, if she finds that her prejudice is deep rooted and psychologically hard to eradicate, she will recognise that because the practice is not wrong she must suppress her prejudice, tolerate the practice, and try to ensure that others and society at large do likewise.

To return to Glover's example, if as a result of genetic mixing there seemed to emerge a weakening in the prohibitions that now protect weaker or less intelligent human beings, then the right response would be to resist this, argue against and fight it. Not conclude that we should not after all have permitted the mixing.

Two points need to be emphasised. The first is that of course we should not give ourselves all this trouble for nothing. There is no point in permitting practices, experiments, or policies that will or may cause huge adverse reaction for little or no expected benefit. The second is that the legitimacy of all these things turns in the first place on whether or not the practices, experiments, or policies can be shown to be morally wrong in some way.

This brings us back to:

Beauty and the Beast

If it is right to try to avoid the suffering caused by deformity or disfigurement can it be wrong to try to select for or engineer physical traits that we happen to admire or like? Should we permit parents to determine things like gender, hair and eye colour, physique, height, etc.?

This is often taken to be a difficult question and indeed the idea of parents being able to choose such things very often causes outrage. Despite my best endeavours I am unable to see this question as problematic. It seems to me to come to this. Either such traits as hair colour, eye colour, gender, and the like are important or they are not. If they are not important why not let people choose? And if they are important, can it be right to leave such important matters to chance?

To this challenge a number of responses might be, and have been made. In the first place we must consider gender. It is often suggested that if parents can choose or are permitted to choose the gender of their children this will lead to a severe imbalance between the genders and a likely superabundance of males, particularly amongst Asian communities. There are two sorts of things that might be wrong with selective bias in favour of one gender or another. The first is sexism. It might plausibly be regarded as sexist if the production of males rather than females is seen as an expression of, say, a preference for males over females and of the idea that being female is somehow a disability to be eradicated if possible. The second is the misery that this would probably cause. The men of the next generation are hardly likely to thank their parents if there is such scarcity of females that the chance of making the same mistake again is radically reduced.

There are of course many non-vicious reasons for gender selection. The obvious ones occur in circumstances in which parents are carriers of sex-linked disorders and where the only way of ruling out such disorders is to avoid children of the gender which may be susceptible. But there are others. For example, where parents want, say, to have children of both genders they may go on trying until they achieve the desired outcome. In circumstances where it is desirable to limit population it is surely much better to let the parents choose and thus limit the size of their family than to leave them to try repeatedly for the daughter or son they so desire.

Indeed, there seems to me to be every reason to permit parents a free choice in matters such as this. Once the notion of a good social reason for having children of a particular gender is recognised it is difficult to see a plausible sticking point. Mary Warnock has defended the idea that the English aristocracy might protect hereditary titles by engineering male successors and if this is not a frivolous reason for gender selection it is difficult to imagine what one might look like! Rather than trying to evaluate various reasons that might be given it seems better simply to recognise the legitimacy of parental choice.[28]

I do not in fact regard the gradual eradication of one gender by these means as likely. A moment's thought should show parents that they would be foolish to produce only children of one gender. And in so far as this foolishness prevailed, the mistake would be likely to be self-correcting. The next generation would put a premium on producing children of the gender required to remove the imbalance.

However, in so far as this sort of discrimination proved likely to be systematic and to constitute not only a threat to the survival of one gender and hence to human beings as such, but also an attack on the dignity and standing of members of the disfavoured gender, we would have reasonable grounds to doubt the morality of such gender discrimination and so have grounds for trying to regulate such behaviour. However, to suppose that gender selection represents a threat to human survival is for all practical purposes quite implausible.

We should bear in mind that having a preference for producing a child of a particular gender by no means necessarily implies discrimination against members of the alternate gender any more perhaps than choosing to marry a co-religionist, a compatriot, or someone of the same race or even class implies discrimination against other religions, nations, races, or classes.[29]

Moreover, the regulation required to preserve a desired and desirable balance in a particular community may fall well short of legislation or prohibition. The tax system could for example be used reasonably to provide incentives and disincentives to behaviour which would fall far short of depriving people of morally defensible choice while at the same time averting what might be perceived to be a catastrophe.

As I have indicated, I have what I hope are well grounded doubts that there is in fact anything wrong with selecting for gender, even in a systematic way which produced a one-gender world, so long as such people could reproduce, avoid other moral wrongs, and so long as they might reasonably be expected to be happy. These are big caveats of course, and since I have examined at length elsewhere the possibility of producing an all-female world and the moral arguments for and against so doing, I will not rehearse them again here.[30]

Perhaps, though there are other dangers. Edward Yoxen, regards one of the main dangers as not gender imbalance but 'the fact that fewer women would be first-born children'.[31] The problem, as he sees it is 'the cumulative psychological effects of general preferences about sex, enacted family by family. One form of this is a passionate desire for a first-born son, or a male only child. Another, which is less worrying, is the desire to balance a small family with at least one of each.' 'The former', Yoxen believes, 'suggests an ingrained male chauvinism, coloured by deep scepticism about what women could ever do, the latter some recognition that the sexes are different, and that both have their value, although males are conceived first.'[32]

A possible danger here, in addition to the sexism that might be involved in parental choice, is the resentment that might be felt by the children. Children of either gender might well resent being 'made to order' and might resent their parents for so acting. However, I am far from convinced that there are any cogent arguments for the entitlement to be protected from feeling resentment and the children in question might find it salutary to reflect that but for their parents' choice, they themselves would not exist at all. For a different selection would mean a different child. Better far to let the children's views feed into the pool of ideas that shape free choices in these matters than to try to pre-empt them by legislation or coercion.

Despite the dangers I sympathize with Yoxen's conclusion: 'Gentle mockery and critique are more defensible and likely to be more productive in the long run. In the case of sex predetermination, too, it is better to retain a degree of procreative freedom, and to argue with people over how they use it, than to deny it to them.'[33] While controls might be exercised via taxation or other incentive schemes the temptation for pervasive and punitive methods of control would be ever present and all too likely to be abused. Even incentive schemes or taxation are likely to be punitive to just those groups most likely to fall foul of them. Moreover, as I have suggested, gender preference is not necessarily an expression of sexism and the task of sifting the motives or the effects of particular selections would be impossible. It is better by far to champion freedom and to fight prejudice by other means.

The right to select for other characteristics that might be genetically engineered into humans or produced by say, careful choice of gametes, is I think straightforward. If it is not wrong to hope for a bouncing, brown-eyed, curly-haired, and bonny baby, can it be wrong deliberately to ensure that one has just such a baby? If it would not be wrong of God or Nature to grant such a wish, can it be wrong to grant it to oneself? It is likely that there will be a sufficient diversity of wishes to ensure that humanity continues to flourish in a myriad of ways, while at the same time reducing barriers to flourishing such as disappointment and disability. If free choice in reproduction begins to look as though it will produce harmful standardization we could of course revue the question of the desirability of controls. No question is ever finally closed.

I am assuming of course that there are no probable adverse side-effects from the procedures necessary to grant the wishes of would-be parents and no reasonably probable hidden dangers. If this is the case I cannot see why people should not be their own fairy god-mothers, at least so far as these morally inert characteristics are concerned.

It is time to turn to a further examination of those characteristics the deliberate production of which might be highly consequential. This will be the subject of the next two chapters.

8

Changing the World

Do we change the world or do we change the people?

The best philosophers from Plato to Marx and beyond have accepted that the task of philosophy is not simply to understand the world but to change it. As Marx remarked (but it might easily have been Plato) 'The philosophers have only interpreted the world, in various ways; the point however, is to change it.'[1] This does not of course mean that we should not try to understand the world. We could not know how or why to change things without understanding them, but Marx's slogan does serve to remind us of the futility of mere understanding without an attempt to apply that understanding to make the world a better place.

Advances in medical molecular engineering have highlighted a particular dilemma. It is that as our power to change people increases, our motive for attempting to change the world may diminish. A second problem concerns the ways in which our power to change the nature of some people, to create a new breed, raises profound questions not only about human nature but about justice.

We must consider carefully both these important issues and they will form the subject of the next chapter. Before examining them, however, we must begin by taking up some loose threads. These threads form a nest of arguments against manipulating the genetic structure of human individuals. The unravelling of this nest and the examination of its various components will form the preface to these deliberations.

The moral difference between somatic line and germ line modifications

We have already considered the difference between modifications that might be made to the somatic cell line of an individual and modifications to the germ line. The difference that interests now is that modifications to the somatic line are alterations which will change the genetic structure of that individual but will not be transmissible to its offspring. Alterations to the germ line on the other hand will change the genetic structure of the individual in ways that will be transmissible indefinitely to its progeny.

A number of recent writers have insisted on the moral importance of this distinction and have argued for a ban or at least a moratorium on germ line manipulation. The Glover Report to the European Commission[2] for example suggests unequivocally that there should be a moratorium on positive human genetic engineering. In stronger terms, the two authors of a new book[3] on the ethics of genetic engineering roundly condemn the idea of germ line manipulation. Suzuki and Knudtson called their book by the rather infelicitous title *Genethics* and it purports to identify nine 'genethical principles', ethical principles which should govern the development of genetic engineering techniques. Their book, as will be already evident, is heavily over-engineered on the linguistic side; it is also rather inadequately comprehending of the ethics. It is one of the few works to date that does, however, have a lot to say about the science and it is this that is its redeeming feature. Of the so-called genethical principles that the authors identify some are hardly plausible as candidates for the status of ethical principles, being either simple platitudes or homilies. The fourth principle is the one which presently concerns us and it is stated thus:

> While genetic manipulation of human somatic cells may lie in the realm of personal choice, tinkering with human germ cells does not. Germ-cell therapy, without the consent of all members of society, ought to be explicitly forbidden.[4]

While rather beside the present point it is important to note the formal defects in the formulation of this principle. By 'formal' I mean those defects which are present on the face of the formulation without the necessity of examining what if any arguments might sustain it.

The first is the odd appeal to personal choice or autonomy. The principle of autonomy gains what force it has from the fact that it is a principle of self-determination which protects the individual's right and capacity to make personal choices and determine her own destiny. But in this case, Suzuki and Knudtson seem to be content to defend the right of someone (anyone?) to determine the genetic constitution of another individual and are doing so in the name of autonomy. This is more usually called paternalism, and while it may be defensible, its defence must lie in the moral reasons for changing the genetic structure of other people, not in terms of some vague waft of language in the direction of free choice.

Then there is the tendentious use of the term 'tinkering'. Of course no one should tinker with the genes of another, but should they repair those genes if they are defective or improve them to the advantage of all? Finally comes the appeal to democracy. In the first place no democratic theory has ever realistically espoused the idea of complete unanimity. If the arguments, the moral reasons for genetic engineering, are compelling is one abstainer to be permitted to sabotage the whole enterprise? On the other hand, if the arguments are equally compelling in the other direction are Suzuki and Knudtson suggesting that the dooming of humankind is acceptable and ethical merely because a consensus will accept it?

We must now turn to the substantive arguments against germ line manipulation. The first is an argument from the utility of genetic defects.

1. The argument from utility

This is of course an argument both for and against germ line therapy. The utilitarian argument in favour of operating on the germ line rather than on the somatic line is this: if someone carries a defective gene they will pass it on to their offspring and so on in every generation. A somatic repair to the defective gene will die with its recipient and his or her offspring will require a repair in their turn. However, if we repair the germ cell, this will be passed on indefinitely and the need to effect repairs in every successive generation will be avoided. This will obviously maximize the good done as well as minimize the costs of doing it.

It might look as though a ban on germ line therapy would appeal only to those who believe doing a little good is morally preferable to doing a lot of good. However, it is often claimed that sickle cell disease provides the knock down argument against such naïve optimism.

The case of sickle cell disease

Sickle cell disease or sickle cell anaemia as it is sometimes called is a rare genetic blood disorder found principally in black populations in

Africa or in populations of African descent in other parts of the world. It is one of those genetic disorders (Tay-Sachs disease is another, affecting as it does primarily European Jews) which affects principally a particular, independently identifiable, group of human beings. It is a painful, debilitating, and life-shortening condition. There is no effective treatment or long-term cure for sickle cell disease but following isolation of the particular gene responsible, gene therapy offers some future prospect of repair and prevention.

The reason that sickle cell disease is interesting to those worried about germ line therapy is that the possession of the sickle cell genotype seems to confer some resistance to malaria. However, it is difficult to know quite what to make of this fact. Suzuki and Knudtson experience just this difficulty:

> The message in selective advantage of the heterozygous sickle-cell trait condition is not that the sickle-cell gene ought to be forever preserved in the human gene pool, like some rare species of endangered animal . . . But the story of sickle-cell anemia does underscore the striking capacity of a seemingly defective gene to simultaneously offer both advantages and disadvantages—depending on its quantities and surroundings.[5]

The message or lesson that Suzuki and Knudtson draw is ecapsulated in these rhetorical questions:

> how many other 'defective' genes responsible for hereditary disorders might harbor some unseen evolutionary value? And, until we know enough about human genetics to begin to grasp their evolutionary roles, what price might we eventually pay if we overzealously try to 'cure' these genetic abnormalities . . . ?[6]

The conclusion we are invited to draw is that we should never risk losing what may prove to be a valuable genetic resource by attempting completely to eliminate a deleterious gene.

Now so far, sickle cell disease is perhaps the only[7] damaging condition known to have a compensating genetic advantage. We do not of course know how many others may prove to be similar but a first question that arises is: how great is this compensating advantage? Wouldn't it be better to try to reduce sickle cell disease as far as possible and continue to fight malaria by the other means available to us—attacking the malaria-carrying mosquitoes and using the highly effective drugs we already possess. The gain in malaria resistance is not as great as the loss to those forced to suffer sickle cell disease.

Of course objectors can say 'yes' to this but still insist that the case of sickle cell disease indicates the reality of unforeseen and unforeseeable dangers in germ line manipulations.

Better the devil you know

Again a general point arises. Any change may have unforeseen and unforeseeable consequences. If we remove, as we have already done, a disease like smallpox from the ecosystem we do not know what effect this distortion of the natural balance will have. Maybe smallpox was useful in containing some other more terrible disease or prevented the evolution of some voracious species? This 'better the devil you know' argument has of course a very general and wide appeal, but it appeals principally to friends of the devil.[8] What weight we give such an argument depends upon the disadvantages of living with a particular devil. If she is tolerable and there is substantial reason to fear a worse alternative, then maybe we should be prudent. But it is not prudence to continue to suffer a terrible condition for fear of a worse which is remote both temporally and in terms of probability.

Are defective genes really 'defective'?

The suggestion of Suzuki and Knudtson is really that there is no such thing as a defective gene, only a temporarily damaging gene. Let's grant this for a moment. Their solution and that of others is to confine gene manipulation to somatic changes alone. It is better, so the argument goes, to face the prospect of making somatic changes in every successive generation than to make a once and for all change to the germ line with unforeseeable consequences.

One problem with this 'solution' is that it would nullify one of the supposed benefits of opting for somatic line rather than germ line therapy. Suzuki and Knutdson did after all insist on asking 'how many other "defective" genes responsible for hereditary disorders might harbor some unseen evolutionary value?' And the point of their question is to remind us that we should ensure that we benefit from this value. But if we permit somatic changes in every generation to eliminate 'defective' genes, then exactly as with germ line therapy, we would never discover any potential benefits which might arise under different environmental conditions. The only way we could attempt to test Suzuki and Knudtson's hypothesis is to leave a few 'guinea pigs' with genetic disorders to suffer for generations while we watch them to see what unforeseen advantage their plight might confer on the rest of us.[9]

Of course, if we could foresee the particular advantage that a 'defective' gene might bestow (resistance to radiation damage perhaps) and could estimate how probably we would require to benefit from this advantage we might have a motive for preserving the defect. But the level of benefit and the probability of our needing it (allowing for solving the problem another way) would have to outweigh substantially the moral costs of condemning individuals to continue with the deleterious effects of the defect.

There is a second feature of this supposed safeguard which makes its benefits far from apparent. If it is realistic to make somatic changes in every generation why not change the germ line knowing that we may have to reverse these changes if things go wrong and remaining alert to this possibility.

At best this would involve a 'one-off' reversal of the change to the germ line and at worse successive somatic changes in every generation to reverse the damage. But this is no worse surely than contemplating the somatic changes in every succeeding generation that were thought tolerable in the first place and which might, of course, never be needed?

Of course all this assumes adequate technology, particularly the ability to target the appropriate gene. And of course if we have reason to suppose that we lack the technology we should not embark on the process in the first place. But adequate technology must be assumed for somatic modifications as well, and the present state of knowledge suggests that germ line modification is likely to prove easier and more effective than somatic manipulation. Indeed, if the change involves the insertion of a completely new gene into the human genome this new gene may be easier to target than modifications to existing genes and hence easier to reverse. We will be returning to this point in due course.[10]

2. The argument against eugenics

A second argument against germ line therapy which is in fact a general argument against genetic manipulation involves raising the spectre of positive eugenics. As Suzuki and Knudtson have put it:

> Underlying the altruistic aims of germ-line therapy lies the potentially dangerous perception that a troublesome gene can somehow deliberately be erased from the human gene pool. The history of genetics suggests that once a human characteristic—such as a particular skin color or mutant hemoglobin molecules or poor performance on I.Q. tests—has been labeled a genetic 'defect,' we can expect voices in society to eventually call for the systematic elimination of those traits in the name of genetic hygiene.[11]

The main argument here seems to be that so nicely ridiculed by F. M. Cornford,[12] 'that a few bad reasons for doing something neutralise all the good reasons for doing it'. The genuinely corrupt and corrupting suggestion implicit in the passage is that to regard some trait as troublesome, defective, or disabling involves the judgement that those who possess the trait are troublesome, defective, or disabled *as persons* in some existential sense and that in consequence they are less than human or less than persons and are themselves eliminable.

There is no logical difference here between the attempt to eliminate a trait by genetic means and the attempts familiar by other means. Take any relatively non-serious but troublesome medical condition, short sight or colour blindness, measles or an active mole: the attempt to rectify, cure, or ameliorate these conditions does not imply that the person who has them is unfit to inhabit the earth. Or consider the desirability of improving something even more clearly multifactorial like intelligence or IQ, which in fact holds little prospect of genetic manipulation; is it somehow evil (or fascist) to wish to improve or even (heaven forbid) attempt to improve the intelligence of our children? And if, as we have asked before, it is not wrong to attempt to improve something like intelligence by education, why should it be wrong to attempt to improve it by genetic manipulation?

Of course this assumes that genetic manipulation of traits or conditions such as these is both possible and safe. If it is neither then it will be this fact that constitutes the objection, not that we are trying to improve the human condition. Any moral being must have this as her goal.

This brings us to the next objections, which are more practical. Again it is convenient to use Suzuki and Knudtson's formulation of these arguments. They produce two complementary objections. The first we can call the argument against gardening.

3. The argument against gardening

This argument follows and takes its departure from a characteristically arresting and resonant warning:

> The notion that the human genome is like some sort of genetic garden from which hereditary defects can simply be plucked like so many weeds is both mistaken and dangerously naive.[13]

Unweeded gardens are characteristically possessed by things rank and gross in nature; we need a strong argument to leave the garden of the soul to such a fate. The argument which supports this suggestion is then presented as follows:

> First any genetic disease can only be provisionally defined. A gene that fails to perform to our satisfaction under one set of nutritional, climatic or other environmental conditions might possibly perform quite satisfactorily in another setting. In this sense a gene's 'defectiveness' can sometimes be a transient quality. And, simply by identifying the nutrients or environmental conditions necessary to sidestep the harmful effects of a mutant gene, one might be able to treat some hereditary disorder without recourse to genetic intervention.[14]

Although this objection occurs under the doom-laden heading of 'the perils of germ-line therapy' this first objection is of course not so much an objection as a reminder that there may be a number of strategies open to us to combat genetic defects or even genetic 'defects'. The moral issue must surely be this; if we are faced with a serious genetic disease which causes pain, disability, or premature death, we should try to cure it if we can. And we should do so by the most effective and expeditious means, for delay will cost lives or human suffering. If there is a gene therapy route available and also what we can call an 'environmental' route then obviously we must compare them carefully.

However, Suzuki and Knudtson's assumption seems to be that the genetically engineered solution should necessarily be the solution of last resort. As yet I see no powerful argument for this, although we have yet to complete the survey of available arguments. Surely, what we should do is compare all possible solutions and choose the best. This of course means the best for us all—the best for the world, not simply the best for those immediately at risk. However, the dangers to those not immediately at risk, including future generations, are not automatically weighted more heavily in the scales of morality than the fate of those presently at risk. Everyone counts for one and none more than one. But real and present dangers have a more urgent claim than less likely and remote dangers. We must all accept that rescues are seldom risk free. The obligation to rescue our fellows,[15] particularly from death or the threat of death, would be non-existent if it could be cast aside wherever and whenever there was some risk involved. We cannot always prefer the duty to future and possible people to that which we have to real and present neighbours. Yet that is what is implied by an automatic preference for environmental solutions. The duty to those more remote from us in time must be discounted for the remoteness of probability of their danger. Although remoteness in time does not always reduce probability it often does and this must be taken into account.

Of course the environmental risks must be given due weight and where they are substantial and highly probable and would cause more harm to future people than avoiding them will cause to present people and where there are no reasonably likely alternative measures of averting this harm, then, and only then, we may rightly sacrifice the lives of real and present people in order not to permit a worse tragedy to occur. But we must be satisfied that this really is the alternative we face and not merely renegue on our obligations to those we could now help because we entertain some vague and perhaps undifferentiated fears about what might conceivably happen in the future and because we hope eventually to discover an alternative remedy.

It might be both more rational and more humane to hope that when and if the damage occurs we will have also progressed to the point where we have other measures of controlling it.

The move to the second consideration takes us out of the unweeded garden of the sacred and fossilized ecosystem and into a very different part of the wood.

Second, there is a continuous upwelling of new disease-causing DNA sequences in the human genome as a result of random mutations. Genetic imperfection is an unavoidable characteristic of human hereditary processes; it is part of what makes us human. And, as the source of both the raw genetic variability on which natural selection depends and the inevitable errors that sometimes result in genetic disease that imperfection is always double edged.[16]

There seem to be two main points. The first is that 'genetic imperfection . . . is part of what makes us human'. And the second is that 'genetic imperfection . . . [is] the source of . . . the raw genetic variability upon which natural selection depends . . .'. What are we to make of these two suggestions?

4. To err genetically is human

In the sense in which to err genetically is human, to err genetically is animal also. All other species are prone to such error and this vulnerability does not of course distinguish humans from other species nor does it mark out anything distinctively human.

Suzuki and Knudtson may here have been taking Shakespeare rather too literally. When Hamlet speaks of the 'thousand natural shocks that flesh is heir to',[17] he is not articulating a universal truth about human nature and the precise number of shocks of which it is comprised—like the precise number of genes in the human genome perhaps—such that if one shock were to be removed we would no longer be human. While it may well be that it is impossible to reduce the number of natural shocks to zero one task for a moral agent is to reduce the number as far as is humanly possible. This is, among other things, the task of medicine. We are not less human because we no longer have to fear the Black Death or smallpox and those of us with an artificially induced immunity to tetanus are not monsters on that account. So it is and will be with genetic immunities. While it may be human to err genetically, it is inhumane to let another person suffer one unnecessary natural shock.

The fallacy here, and for once it is proper to talk of something as hard and concrete as a fallacy, is that human nature is constituted by its complete description at a particular moment in time. In other words that human nature just is the nature of the humans now existing. Human nature is changing and evolving constantly and we are very different from our ancestors. Our descendants, if the species survives, will differ from us in ways it would be hard to predict. We have changed and can still change radically and still be human. What is implied here of course is that there are some characteristics which are normatively human, that is, characteristics which we want humans to have. What must be said here is that not all wants are of equal moral status. Some of the things we want are evil or harmful. We may be entitled to satisfy such wants for ourselves, what we are not entitled to do is impose the consequences of such desires on others.

It is hard to believe that any person would want others to be so constituted that they suffer and perhaps die prematurely when there is an alternative available. What is clear is that such a preference cannot be a moral preference nor can it form part of or be the obvious consequence of an ethical principle.[18]

5. The sword of diversity

The preference for diversity is ancient and healthy. In espousing it so strongly Suzuki and Knudtson are placing themselves in a long tradition of respect for the multiplicity of human flourishing. The problem however is that they are misplacing themselves in this tradition.

Their eighth 'genethic' principle is:

> Genetic diversity, in both human and nonhuman species, is a precious planetary resource, and it is in our best interests to monitor and preserve that diversity.[19]

One can only say 'right on' to such a principle, but so saying does not bind us to reject gene therapy nor even to forswear manipulating the human germ line. There is an equivocation here as to the meaning of 'diversity'. Genetic diversity is simply the diversity of types of any one gene. The importance of genetic diversity is that it is believed to confer the capacity to resist damage and disease on the one hand and confer adaptability to the organism on the other. Human diversity by contrast is something like 'variability of phenotype' and it results from the interaction of genes one with another, and on genetic interaction with the environment. There are an estimated 100,000 genes in the human genome comprised of some 3,500 million base pairs each of which can make a difference.

Even deleting one or two, or even ten or twenty defective genes in one individual or altering the same number of base pairs[20] is not going to make a significant contribution to either genetic diversity or human diversity or lack of diversity. And we must recall that some modifications (perhaps most) will not require deletion of a gene but repair or modification of a gene or perhaps even, as we have seen, insertion of a completely new and additional gene.

At the moment the technology is relatively crude and it is more likely that the whole gene or large chunks of DNA comprising perhaps thousands of base pairs will be deleted or replaced in gene therapy. However, as techniques become more refined, smaller segments of DNA may more standardly be targeted for repair, deletion, or substitution.

But even the worst case scenario from the point of view of gene deletion will only affect an infinitesimally small proportion of an individual's genes. Moreover, for the foreseeable future the probable numbers of engineered individuals is so infinitesimally small a proportion of the human population as a whole that there can be no serious threat to human diversity. But even if we do eventually manage to make the advantages of genetically engineered improvements available to all there is no reason to fear a diminution in diversity. Indeed, genetic engineering may even increase diversity. And of course we must not assume that we will not be able 'artificially' to increase diversity by engineered means in the future to redress any imbalance that might, *per impossibile*, have occurred.

6. The meaning of diversity

We must also bear in mind the different concepts of diversity, genetic and human, that we have noted. When the spectre of reduction of diversity is raised we are invited to think of the term 'diversity' as it is commonly understood, that is as human diversity, rather that in its more restricted sense of genetic diversity. This, as it is generally understood, refers to things like: what people look like and what they can do. The former is not likely to change radically and the latter is likely to be increased by gene therapy rather than reduced. This is because genetic defects usually reduce individuals' capacities in two ways. First by restricting them physically and mentally and secondly by shortening their lives. If more people have longer and less restricted lives this is likely to maximize human diversity rather than minimize it. Of course, Suzuki and Knudtson may not be trying to raise this spectre, but then the force of their objection or the rationality of their fears are lessened. We must remember that genetic diversity narrowly understood can only be good for two things. Either it enhances human diversity in the first, common or garden, sense; or it maximizes human adaptability and the chances of human beings meeting new dangers of various sorts. As I have argued, it is implausible to believe that removing or modifying the genes that cause disease will in fact alter human genetic variety in any way at all, let alone in any significant way and this conclusion is borne out by the final arguments of Suzuki and Knudtson, which we will consider in the next section.

Moreover, it seems more than a little perverse to insist that we humans should remain susceptible to genetic disease now on the off chance that this present vulnerability might somehow reduce our susceptibility at some future time.

7. It's impossible anyway!

Finally Suzuki and Knudtson produce a group of objections all of which seem to say that genetic engineering is impossible anyway or at least impossible on a useful scale.

> [T]o try to eliminate a . . . gene causing a rare genetic disorder, one would have to treat not only the occasional . . . individuals who suffer from the disease but also . . . carriers. This would require massive genetic screening programs to sift through the genotypes of the entire human population, the vast majority of whom were healthy. A less ambitious program aimed strictly at repairing the genes of . . . individuals (at risk) and thereby culling their defective alleles (genes bearing hereditary characteristics) from the gene pool presents other problems. The decline in the frequency of the defective gene would proceed at a ponderously slow rate. As the frequency declines, however, the cost of diagnosing each case would show a corresponding rise. So inefficient would such a negative eugenics program be that it might take hundreds of generations and many centuries to achieve a meaningful decrease in the frequency of a deleterious recessive gene.[21]

Of course problems like this are real and important, although I doubt they are as decisive as Suzuki and Knudtson imagine. We may be on the verge of an era in which each human individual has her genome mapped routinely *in utero*, at least in the developed world. If this became standard antenatal practice the task of screening the human population would gradually reduce although of course such high-tech antenatal care would not be available world-wide for the foreseeable future.

However, the difficulties of genetic screening on a total population basis are not objections, let alone ethical objections to gene therapy as such. Quite why Suzuki and Knudtson are so gleeful about them I am not sure. They occur in the section entitled 'the perils of germ-line therapy' and it is an obvious relief to the authors that they feel able to argue that the process which they regard as so perilous, is also so impractical.

The fact that we cannot offer something to everyone does not constitute a reason for not offering it to anyone, particularly when what is on offer is protection from severe disease. We should therefore regret rather than celebrate the difficulties in the way of genetic screening. If and when such screening, perhaps in the form of mapping the genome, becomes available one question is whether it should be introduced as routine antenatal care wherever possible. Cost permitting, the answer to this question must surely be 'yes'.

8. Multifactorial problems

Finally and perhaps most important, the overwhelming majority of inherited human traits—intelligence, height, and skin pigmentation, for instance—tend to be polygenic. That is, they arise from the interplay of more than one gene . . . For now, at least, such multi-gene targets remain beyond the reach of those who might consider employing new gene therapy techniques as a tool to shape their particular dreams of human genetic hygiene . . . Even assuming [the complex repairs necessary] could be carried out, one could never be sure that altering a constellation of polygenes would not inadvertently disturb other, seemingly unrelated cellular processes influenced by these genes.[22]

Again whether or not it might be ethical to attempt to alter a 'constellation of polygenes' would depend on what was at stake and what a reasonable assessment of the risks revealed. It would be hardly credible to suppose that it would be worth running even the slightest risk to achieve a change in pigmentation for its own sake. However, there might be polygenic traits which it was highly important to attempt to change even at some slight risk. We surely would not want to rule this out altogether, and indeed pigmentation might even prove to be one of these.

Suppose, say, as a result of ozone depletion it proved essential for human survival to change skin pigmentation in order to resist the ill effects (perhaps cancers) of solar radiation. In such an event we might well regret rather than celebrate the polygenic nature of the trait involved and might well regard running some remote risks as acceptable given the alternative.

9. The obligation to future generations

It is difficult to know quite what to make of the claims of future generations or even to know just how to characterize them. Suzuki and Knudtson talk in just these terms. Noting the moral difference between somatic cells and germ cells, they talk of these latter cells 'with their potentially immortal genes to which future generations also lay claim'.[23]

Talk of the claims of future generations is slightly awkward for it seems to imply that these generations already exist in some sense and are presently making claims upon us. It seems happier to think in terms of our obligations to future generations. These are of two kinds. The first kind is not so much the question of our obligations, if any, to those who may exist in the future but rather the question of whether we have any obligation to ensure that there will in fact be future generations. The second deals of course with the question of what, if any, specific obligations we have to those who will come after us.

The obligation to ensure that there will be a future

Again there are different ways of thinking of this obligation. We might think of it in terms of the obligation not to destroy the earth or the obligation to preserve the ecosystem. Or, we might think of it in terms of the obligation to reproduce our own species. These are distinctly different because whereas if we think we have no obligation not to destroy the world we must, *a fortiori*, accept that there is no obligation to continue the species, the reverse is not true. We might judge that there is no particular reason for human beings to go on reproducing themselves but still think that we should not be spoil-sports and leave nothing for the other creatures who inhabit this particular world.

I shall not take time here to argue the point, but it seems to me clearly wrong for human beings to decide to destroy the world. For one thing, this would almost certainly, in any imaginable scenario, involve the wilful murder of presently existing persons. But even if all agreed (and I leave aside the problem of very young children and others incapable of autonomous choice) it would still be wrong to rid the universe of something so complex and beautiful, and moreover something which, perhaps uniquely in the universe, is capable of sustaining sentient life.

However, the wrong of all presently existing individuals, say, simply deciding not to reproduce, simply deciding that the present generation should be the last, is of a different order. It is different because it would not involve violating the will to live of any person, nor the destruction of the ecosystem. Indeed there is a sense in which it might be better for the ecosystem as a whole if this were to happen—better in the sense that the system (humans apart) would be less in danger of total destruction or more limited damage from environmental pollutants, more varied, and more likely to flourish. However, it would I think be wrong for two distinct and important reasons. The first is that it would be to prefer a universe with less happiness and less satisfaction of desires than the alternative in which persons did continue to exist, and secondly because it might involve the permanent end of the only creatures anywhere who have both these capacities.

These points all raise difficulties which could be discussed at considerable length, I mention them here merely to note them. Our present concern is with the second sort of obligation, and more particularly with the problem of what if any obligations we have to future persons on the assumption that there will in fact be future persons?

The obligations we have to future generations

Again there are many specific sorts of obligations we might think of in connection with future generations, obligations not to spoil the environment for them or use up valuable natural resources or leave behind sources of danger like spent nuclear fuel, unexploded nuclear weapons, or other dangerous pollutants. However, it is convenient to think of all these specific obligations we might have towards future people as obligations not to harm them.

There are two standard ways of bringing about harm. One is by using what have been called 'positive actions' and deliberately changing things so that harm results. The other equally effective way is by deliberately leaving things as they are knowing that harm will result—using so called 'negative actions'.[24] This idea is sometimes called the 'acts and omissions doctrine', the doctrine that there is no moral difference between acts and omissions with the same consequences.

Although once controversial and still often opposed this view now seems incontrovertible.[25] It would obviously be as bad, say, to pollute the environment so that the incidence of cancer increased by 20 per cent as it would be not to remove a naturally occurring environmental hazard when we could easily and when failure to remove it would cause the same degree of harm. Or, remembering the example in Chapter 3, it would be as bad for a mother to fail to take a drug during pregnancy that would remove a disabling condition from her child as it would be deliberately to take a drug that would bring about the disability.

Our obligation not to cause harm to future generations has the same positive and negative faces. We must not act positively so as to cause harm to those who come after us, but we must also not fail to remove dangers which, if left in place, will cause harm to future people. Thought of in this light there is a clear dilemma about genetic engineering. On the one hand we must not make changes to the genetic structure of persons which will adversely affect their descendants. On the other hand we must not fail to remove genetic damage which we could remove and which, if left in place, will cause harm to future people.

We must in short weigh up the probability of harm occurring as a result of what we do, against the probability of harm occurring if we fail to take steps now to prevent its future occurrence. In some cases the dilemma may be acute, we may simply not know enough to be able to make reasonable judgements as to the various probabilities involved. In such cases we should err on the safe side. It is a singular fact that the safe side is always supposed to be the side of inaction or preserving the status quo. I am not sure whether this assumption is sound, but if it is, then we should avoid, for example, modifying the germ line if we have good reason to suppose that there may be adverse effects.

However, we should not fail to remove a danger which is real, present, and certain for fear that there may be some adverse consequence which we have no reason to expect but which might conceivably occur.

Caution or negligence

The important lesson is that we do not avoid trouble, nor yet the possibility of being responsible for harm, by doing nothing. Doing nothing is not necessarily the safe option. Whether or not it is, depends upon careful consideration of the probable consequences of the alternatives. Sure, we should be cautious, but we should not be negligent. And it would surely be negligent to deny human beings the benefits and protections that gene therapy will surely make possible.

Aborting Beethoven

In argument about genetic engineering when the question of screening for heritable disabilities is raised the following story is always told: you are advising a pregnant mother and tests show that the child she is carrying is suffering from inherited syphilis and is highly likely to develop an associated deafness. Would you recommend a termination? This story is always told to a stooge who is supposed to reply 'yes', to which the triumphant response is: 'You have just aborted Beethoven.' Even the redoubtable George Steiner has been seduced by the desire to produce this particular jack-in-the-box. In a 1989 television programme on genetic engineering he said:[26]

> It turns out that what in many cases is a hideous disease, a handicap can also be profoundly creative. Without the kind of meningital deafness which comes of inherited syphilis and alcoholism you and I would be sitting here without Ludwig van Beethoven. Now that to me is absolutely key.
>
> Much of what has been the deepest, the most joyful, let me underline that the most joyful in human thought and creation has sprung out of very profound physical but also mental handicaps. I am not prepared to say that it would have been very much better never to engender the Muscular Dystrophy, if it was that, . . . to which we owe Toulouse Lautrec.

The power of the argument from aborting Beethoven seems to lie in this sort of logic: that aborting Beethoven can only seem a good thing to do if we, or the world, or his family, would have been better off without him. And since this seems an implausible thing to suppose we seem to be forced to the conclusion that Beethoven should not have been aborted and so neither should other fetuses in related circumstances.

But, to believe it right to abort a fetus is not to be necessarily committed to the view that the world would be better off without that individual, nor that the individual would eventually wish she had never been born, nor that that individual will be unhappy, nor that the individual will suffer. Nor in aborting an individual with Beethoven's syndrome are we in any sense aborting Beethoven or a potential Beethoven. In all these cases what we are aborting is an actual fetus and the rights and wrongs of that are settled by a consideration of the moral status of the fetus.[27] The fetus we abort will never become anything, and it is nothing but a fetus at the time it is aborted. It is as senseless to bemoan its loss as the loss of a Beethoven as it is to celebrate its loss as the pre-empting of a Hitler.

However, there is one other disturbing dimension to Steiner's ode to joy. It is the monumentally selfish and self-indulgent preoccupation with his own (and to be fair, humankind's) joy at the expense of others. 'You and I', he says, 'would be sitting here without Ludwig van Beethoven', and this, he seems to believe, would be a tragedy even if it were very costly in terms of human pain and suffering. It is clear that Steiner is really committed to something like this, for the passage quoted above continues:

> We know very little about how much pain and suffering are positive in equations . . . of human dignity, of human decency of human altruism . . . somebody who has to be up all night with an incontinent relative or someone screaming in pain of Altzheimer's Disease. To say to such a person we will not try and take your burden away is a hideous impudence, at the same time to say in a kind of Aldous Huxley way we are going to create a new sanitised essentially pain free, essentially beauteous set of worlds may be to alter the balance of man's moral dignity, of man's religious questioning far out of any proportion.[28]

Despite the equivocation of this passage it is clear that Steiner is inclined to prefer to rescue the paintings rather than the people if the museum is on fire. He seems to believe that it is better to have a world with muscular dystrophy and Toulouse Lautrec's paintings, than a world without both.

But of course this isn't quite the possibility envisaged by the genetically engineered eradication of genetic defects. The Toulouse Lautrecs and the Ninth Symphony will continue to give joy. Steiner is supposing that to eradicate genetic defects is somehow to cut off the wellspring of genius.

> We are playing with what you call on the Stock Exchange with that haunting word 'futures', and futures can stretch in this case into the centuries, perhaps into the millennia. Who are we to cut off what have been the sources often of our eminence, for a Nietzsche, for a Dostoevsky, for a Pascal . . . how many, what are the units, the Benthamite Utilitarian units of sane good healths—it may be that they outweigh genius.[29]

Again it seems clear that Steiner does not think so, but this is not the point. There is no evidence so far as I am aware that genius is solely or even principally drawn from the ranks of people with genetic disorders, nor that genetic disorders play a causal role in the genesis of genius. There is equally no reason to suppose that in reducing human susceptibility to genetic disability we will, *pari passu*, be reducing our capacity for genius; that has to presuppose that geniuses are only going to be thrown up by genetic defects and, as I say, there is no evidence for that at all. There is then no reason to think that in reducing the sum total of human misery we will be reducing the sum total of human genius.

A related argument was produced by Germaine Greer in the course of her participation in the same television programme. Just as George Steiner wants to insist that there may be a positive side to pain and suffering which would justify our declining to eradicate it when we had the chance, so Germaine Greer believes that humankind should have the advantage of the disadvantages of particular people.

> Curing a disease is one thing, stopping it existing is another ... people also have the advantage of their disadvantages. If a trait is totally maladaptive then it doesn't survive. If Huntington's Chorea has survived in ... special populations there's a reason for it and we just simply haven't found it ... When we understand more about it we might be able to deal with the deleterious activity of the gene and preserve the positive aspect of the gene.[30]

Well, we might, but at what price? First it is simply an article of faith on Germaine Greer's part that there is a positive side to Huntington's chorea, that it does have some hidden major utility to humankind or maybe to the ecosystem abstractly conceived. I know of no reason to suppose that Greer's claim is true in the sense that all traits or all genetic defects or even perhaps all diseases (why not?) are adaptive in the positive sense of being good for humankind or good for the ecosystem. They may simply for example be good for the destructive trait, organism, virus, or whatever, in question. This is especially obvious in the light of the fact that Huntington's chorea manifests itself only after the peak reproductive period is over.

But more important what price should we put upon the outside chance of Greer being right? Well if it costs nothing to bet on Greer and wait and see then we should certainly do so. But of course it does not cost nothing. It costs her nothing, but it costs those who have and who, *ex hypothesi*, will continue to have Huntington's chorea and other terrible genetic disorders a great deal in terms of pain, suffering, and premature death.

I suggest that we would have to be very confident indeed of a number of things before we might be justified in accepting Greer's cruel gamble. The first is that there was indeed a positive side to Huntington's chorea, second that such a positive side was sufficiently important to be worth preserving at the terrible cost in human suffering that would be required to preserve it, and finally that such positive effects could not be or were not likely to be achievable another way.

In the absence of such confidence I hope that no sane, let alone moral, being would think it worth preserving disability and disease on the off chance that some good might come of it and some unspecified and unpredictable point in the future.

It would be better if we could end this chapter on a sweet note and what better one than the twelve perfumes of rice.

The twelve perfumes of rice

It is again Germaine Greer who gives us this fascinating footnote to the possible evils of genetic engineering. She is not so much interested, as was Steiner, in the idea of genius, but she presents a related argument in defence of the rather more nebulous[31] ideas of 'self-hood' and mythology. But as with Steiner, Greer is again prepared to condemn people to misery rather than risk the loss of certain other features of human existence.

> . . . they bred the taste out of rice and the whole civilization that . . . could name the twelve perfumes of rice was completely disenfranchized, I mean their mythology and their self-hood was gone. They were supposed to eat this miracle starch which would keep them alive, but what for when the twelve perfumes of rice were gone?[32]

The argument of Greer's suggestion is clearly that it is better that people starve than that they eat unperfumed (or perhaps minimally perfumed) rice. While there may be something to be said in favour of such a view, I cannot myself imagine what it might be.

9

The New Breed

Now we must again entertain conjecture of a time when technology permits the introduction of new or modified genes into the human genome. It is time to take further the resolution of problems associated with the idea of improving human beings. We shall suppose the possibility of introducing genes coding for antibodies to major infections, or coding for repair enzymes which would correct the most frequently occurring defects caused by radiation damage, or which might remove predispositions to heart disease, or which would destroy carcinogens or maybe permit human beings to tolerate other environmental pollutants.[1] These techniques will also eventually enable us to insert genes which could repair DNA and have the effect of retarding the ageing process.[2]

In Chapter 7 we examined the problems associated with doing this either via hybridization or in a way which changed human nature. Our concerns then were in a sense personal. We were worried about the effect this would have on the individuals themselves. We must now think more in terms of the effects on society and indeed on the world at large. Clearly the sorts of modifications to persons we are thinking about here represent different possibilities but in large part the ethical problems they bring with them are related. Before examining these, however, we must now sketch in a little more fully what each possibility might involve.

In the case of genes coded for antibodies to major infections, the effect would be that a new gene or genes would be inserted into the human genome which would effectively 'immunize' the individual against the infection in question. We will imagine that the infections include AIDS, malaria, and hepatitis B.[3] It is likely that the gene would have to be inserted in the early embryo before implantation. This would in effect mean that any individuals to be protected in this way would have to be conceived as part of an IVF programme or in a clinic able to use 'lavage'. It is possible that the gene could be inserted surgically, or via a catheter inserted into the fallopian tube through the uterus, but this would presuppose very early detection of pregnancy. Another possibility might be to insert the gene at the gametes stage by introduction either into the egg or into the sperm prior to fertilization. The advantage of introduction into the gametes would of course be that if the male gametes alone could be used then there would be no need for IVF or for an invasive procedure to gather or operate on the egg, and only minimal technology would be required for the necessary artificial insemination to bring the modified male gametes into contact with the egg.

The disadvantage of operating on the male gametes alone would be the difficulty of the procedure. A dominant gene would need to be introduced into the sperm and the chances of passing this on to the next generation would depend on the success rate of introducing the DNA into the sperm. Assuming that modified sperm were not at any disadvantage in fertilizing the egg and that the DNA had sucessfully been introduced into the relevant sperm (the sperm which fertilised the egg) then it would be possible to achieve 100 per cent success in transmitting the gene, but only to the next generation. Thereafter the success rate would vary according to a range of factors but at best it would be a 50 per cent chance of transmission.

To make probable a 100 per cent success rate in transmission of the new gene, via the gametes generation after generation indefinitely, the mature gametes of both parents would have to have been genetically modified by insertion of the new gene into both chromosomes. Since this process is always likely to be problematic, the success of the insertion would have to be checked before conception.[4] Thus, for complete success in operating on the germ line we would have either to modify both sets of gametes, egg and sperm, or insert the new gene into the early embryo. In either case the result must be that the new gene appears on both sets of chromosomes in the resulting offspring.

I. The New Breed

Individuals whose genome had been genetically modified in this way would become what might reasonably be termed a 'new breed'.[5] This term, while doubtless controversial, is I think fair because the modified individuals would indeed be both 'new' and a 'breed', first in the sense that they would have what is an entirely new and unprecedented genetic constitution and also in the sense that their genetic constitution would differ systematically from other human individuals. Their claim to be a breed, though of course they might not themselves wish to make any of these claims, rests clearly but decisively on the fact that they could pass on this new constitution by normal reproduction with other members of the same 'breed'.[6] From now on, when I refer to the 'new breed', I mean to refer to the group of individuals who are genetically modified in this way.

Of course when we here talk of 'parents' we mean 'genetic parents', which need of course only be egg and sperm donors or perhaps embryo donors. It would be possible for the genetically modified individuals to marry or have sexual partners outside the group and reproduce successfully using these sorts of techniques. This factor may have an importance which we will discuss in due course. In the case of genetic measures to protect against carcinogens or environmental pollutants, the techniques would in all probability be the same,[7] as they would with DNA repair to retard the ageing process.

In the previous chapter we considered a set of arguments against embarking on genetic modifications of these sorts and found them all wanting. The benefits of protections against major infections, or cancers, or heart disease, or environmental pollutants and in favour of retarding the ageing process are so great that none of the arguments so far examined give us good reasons to forgo these advantages. But what would it really be like to embark upon a programme of genetic modification of the human species?

This is the question that has been with us from the start, we began to examine it when we considered the desirability of creating wonderwoman and superman or their more modest counterparts. But at that time we were interested more in the specific problem of creating hybrids and in possible reactions to hybrids. The scenario we began to examine then is still unresolved. For whereas we have looked at a wide range of specific arguments, and the fears and doubts that surround ideas concerning 'human nature', we have not yet faced up to the consequences of the deficiencies of those arguments, the groundlessness of those fears, or the insubstantiality of those doubts about the genetic modification of human beings.

A breed apart?

In considering the production of wonderwoman in Chapter 7 we imagined, for the sake of an argument, that we might genetically enhance intelligence. We noted then that intelligence is multifactorial and so is unlikely to prove amenable to genetic modification. For this reason we will assume that intelligence will remain beyond the reach of genetic manipulation and we shall assume that the new breed will have to make do with whatever intelligence they have, whether heritable or not.

We will, however, assume an optimal scenario for the other sorts of genetic modifications we might make to the human genome. Optimal in the sense that we will make it as easy for ourselves as possible. We will not assume, however, that we can necessarily make all the genetic modifications we would wish at the gametes stage, before conception, and so avoid completely the necessity for IVF.

The argument to follow will assume that we have made the breakthroughs necessary to achieve a number of different sets of modifications. We will assume we can insert new genes coded for antibodies to major infections including AIDS, hepatitis B, and malaria. We will also assume that we can insert genes coded for enzymes which will repair damage from carcinogens or environmental pollutants of various kinds and others which might remove predispositions to heart disease. Finally we will assume that we can insert genes which will repair DNA and so, amoung other things, retard ageing.

We need not imagine that we will be capable of preventing all cancers in this way, nor that members of the new breed will be proof against all environmental pollutants, nor that they will be immune to premature vascular disease. We will simply imagine that as well as immunity to specific infections, they will enjoy substantially reduced risk of contracting cancer or heart disease (though they may of course still be vulnerable to some cancers) and that they will be at substantially reduced risk of damage (including cancer damage) from environmental pollutants of various kinds.

Finally we will suppose that members of the new breed can expect to live substantially longer than the normal life expectancy of present day human beings—let's say, for the sake of the argument, that a balanced lifespan will be increased by twenty years on average—that is twenty extra years, neither of youth nor senescence, but balanced over the whole lifespan.[8]

In our scenario, the work has been done, it has been exhaustively tested in animal models and on early human embryos, and the results have been replicated in laboratories in Europe and the United States. There is general agreement that the procedures are risk free and will have the advantages outlined.

Normally one would expect a pilot project studying carefully the first cohorts of human subjects throughout their lives to check for any unexpected side-effects or problems. However, in view of the very long periods of time involved for the pilot study and the immense advantages from the procedure, including an obvious and pressing need to combat AIDS (which we will assume is still without a vaccine or other cure), it is decided to offer the genetic modification as widely as possible to volunteer parents. Needless to say all members of the new breed will be carefully monitored and followed up as if they were members of a pilot study.

Well of course there is no shortage of volunteers. Indeed, the clinics capable of offering the service are flooded with demands and rationing is immediately essential even though the government have made the offering of the new genetic immunities a major priority.

As the programme of producing members of the new breed accelerates in the industrialized world a number of further advantages and further problems emerge. We will list these now and then examine them in more detail.

1. Advantages

1. The creation of the new breed is seen as the most powerful public health measure yet devised. In creating immunities and protections for its members it reduces the 'at risk' population and will inevitably free medical resources for normal humans.

2. As the new breed multiplies a range of health costs will gradually reduce. We can of course expect other problems to take up any slack.

3. The advent of the new breed is attractive to employers in that new breed employees will be less susceptible to disease and presumably less prone to absenteeism.

4. Professions requiring a long training and high investment in training, and so looking towards long service for an adequate return on investment, will be particularly keen to recruit members of the new

breed and to offer incentives. The medical profession might be a prime example here.

5. The new breed will be less susceptible to many occupational health risks, carrying as they do protections against environmental pollutants and cancer risks.

6. Important among such occupational health risks to which the new breed will be immune will be those associated with health care. Immunity to AIDS and to hepatitis B, being obvious examples.

7. Areas which demonstrate higher than normal health risks could be target populations for priority in genetic protection. For example communities with abnormal rates of leukaemia in children might benefit especially from genetic protection. Equally geographical areas, like parts of Cornwall where there is naturally occurring radioactive gas, may also benefit from special priority in the allocation of resources to the genetic protection of inhabitants.

8. Recruits from the new breed would be especially attractive to employers involved in the nuclear industry or maybe in the defence industry or the armed services, who might not only wish to recruit protected employees but also offer existing employees the chance to protect future children.

The list of advantages could continue but we have already passed the point where the 'advantages' are exhibiting their disadvantageous side and we need now to identify precisely some of the drawbacks.

2. Disadvantages

1. Inevitably as individuals protected against environmental pollutants multiply the perception of the urgency of need to eradicate such pollutants might well recede. The existence of the new breed might thus carry dangers to the environment as a whole as well as to the rest of humankind.

2. In providing a potentially protected or even partially protected work-force, the existence of the new breed may decrease pressure to eradicate dangerous working conditions, including and in particular those in the nuclear or chemical industries. Although not dangerous or at any rate minimally dangerous to the new breed, these industries and others like them will continue to present a danger to the world at large, to all its other life forms, and to the majority of human beings who will remain unprotected.

3. Members of the new breed will be perceived by the rest of the human community as not only having an (unfair?) advantage in protection from danger but in consequence an (unfair?) advantage in employment opportunities.

4. Equally, since for the foreseeable future members of the new breed will be drawn largely from the populations of industrialized and technologically sophisticated nations, and also, for reasons we will consider below, probably from the higher social classes of those nations; such nations will be perceived as increasing their (unfair?) advantages over the rest of the world and the 'higher' classes as doing likewise.

5. And in so far as disadvantages 1 and 2 apply, the rest of the world will see the nations capable of introducing the new breed as presenting an environmental danger to the rest of the world and its unprotected life of all sorts.[9]

6. In order to maximise the advantage of their constituting a 'breed', members of the new breed would need to confine their procreational activities to one another. This fact would inevitably lead to some pressure for intermarriage and while the new breed were relatively few in numbers this would create severe problems of choice as well as indicating the desirability of an attempt to balance gender numbers at the stage of genetic modification.[10]

7. I have so far deliberately ignored the supposed biological disadvantages of tampering with the genetic constitution of individuals. Many of these we rehearsed in Chapter 8. For example the possibility that the new gene or genes might disrupt other genes, perhaps by actually being inserted into a gene, or might disrupt the controlling sequences of genes, cannot, at present, be ruled out. I am assuming throughout this discussion, however, that we will be able to target these new gene insertions accurately into a 'harmless' position in the genome. If we cannot be sure of this of course, then the dangers of the procedure will probably outweigh the advantages.[11]

8. Increased global population due to additional longevity of the new breed depending on how populous the breed becomes.[12]

The problem that the balance of advantages and disadvantages presents is of course simply that of deciding which set of options are the right ones. And this is why we have spent so much time in this book dealing with possible objections of one sort or another. The advantages of producing the new breed speak for themselves, they tend to show us that it is the right thing to do. In making sure that it is right we must now consider the disadvantages.

We will come last to those disadvantages which make reference to justice since in a sense, they are the most intractable. First we shall look at the problem of coercion.

Coercion

How great a problem this disadvantage—the last in our list, but the first to be examined—will prove is hard to gauge. We need not imagine any *compulsion* to reproduce only with fellow members of the breed. There may well, however, be social and indeed peer group pressure of the sort used, to a greater or lesser extent, in almost every culture, race, and religion to protect the group or guarantee its continued existence or expansion. Of course, attitudes to the necessity for this or to the respectability of it will vary. We must remember that there is currently related pressure on those carrying genetic diseases not to reproduce with partners whom they know also carry the defective gene where this will lead to genetically damaged children. However, while it is desirable from the social and economic point of view that members of the new breed procreate with one another, this is not of the level of importance to justify anything stronger than the mere observation that the benefits of genetic modification will be maximized the more this happens. And of course I mean no more than just this when I talk of desirability from the social and economic point of view.

Some, perhaps many, members of the new breed will wish to resist such pressure and choose for themselves. And, like the more familiar sorts of pressure we have considered, they are perfectly entitled to resist this pressure if they so choose. Of course in particular circumstances the difficulties in the way of resisting such pressure may be formidable. Others might compromise and choose partners from the community at large but avail themselves of the reproductive technology to replace their partner's genetic contribution with that of a member of the new breed. We can imagine new breed egg and embryo banks, coupled with a sperm donation service, would be set up to fill this need. Of course it would be just as effective if members of the new breed contemplating 'marrying out' were appropriately circumspect and had any children genetically modified. However, in view of the scarcity of this resource they might find that they were not in fact able to secure this advantage automatically for their children, particularly since this might not be a cost effective use of resources. After all, if members of the breed procreate with non-members they will, assuming the trait to be dominant homozygous (and no sensible scientist would try to introduce anything else), still protect all their offspring in the first generation. The problem arises for subsequent generations when the chances of continued membership of the new breed will be at best 50 per cent.[13] In view of this the argument that scarce resources should be devoted to creating new members *ab initio* might be powerful and might be part of the pressure put on members of the new breed to be economical with their reproductive capacity.

We must now consider the outstanding disadvantages. Let's look at what we may term the 'environmental dangers' next.

Environmental dangers

We noted that as individuals protected against environmental pollutants multiply, the perception of the urgency of the need to eradicate such dangers might well recede. Equally, if a protected workforce could be supplied in the shape of the new breed, their very existence might decrease pressure to eradicate dangerous working conditions.

The first danger does not bear close examination. For the foreseeable future, for hundreds of years probably, the new breed will constitute a tiny proportion of the world's population and so the moral imperative to protect people from pollution will remain. We must also remember that people are not the only inhabitants of the planet and the rest of the ecosystem will also require protection against environmental pollutants.[14] Indeed, justice would require that we take even greater steps to see that non-members of the new breed were not further disadvantaged by their lack of protections that members enjoyed.

The second supposed disadvantage of the new breed has slightly more merit in it. It is of course true that if the new breed are indeed proof against dangerous working conditions of particular sorts, then for them the conditions are simply not dangerous. But even if acceptable ways could be found to ensure that only members of the new breed worked in these dangerous places, this would go no way at all to providing a solution to the problem of industries producing dangerous pollutants or radiation. Such industries do not only or even principally present a danger to their own workforce; as Bhopal, Flixton, Sellafield,[15] Three Mile Island and numerous other catastrophes have shown. For this reason no sane person would think the existence of the new breed constituted a reason for being complacent about environmentally dangerous industrial processes.

It is for these reasons that there is no merit either in the suggestion that the existence of the new breed initially predominantly in industrialized societies presents a danger to the rest of the world in that it provides industrialized societies with reasons to be complacent about pollution.

We have now come to a set of problems presented by the possibility of creating the new breed which all have to do with the justice of so doing.

II. Justice

Many of the problems that arise concerning the ethics of producing the new breed turn at some point on the idea of justice. For not everyone, even in the future, will be a member of the new breed, nor will they be able to secure the benefits of membership of the new breed for their children. And where some people continue to suffer disease and premature death while others have been protected against these evils in many cases, the question of the fairness of the fate of the victims, or the fortune of the protected, inevitably arises. There will then inexorably be a problem concerning the just allocation of benefits.

Justice is a notoriously difficult concept of which to give an account. The majority of the most influential writers on contemporary social and political theory have had a go at it.[16] For reasons that are not germane to our present discussion I think that the concept of justice is coherent only when seen as a dimension of a more fundamental moral principle—that of equality.[17] Now the principle of equality is usefully ambiguous and while most people accept equality as a value, many differ as to precisely what this value actually amounts to. I want to suggest the following as a sort of irreducible minimum concept of equality. It is at any rate the one which best explains the imperative for a fair distribution of resources of the sort involved in creating the new breed.

The principle of equality

The principle of equality just is, I believe, the principle that people's lives and fundamental interests are of equal importance and that they must in consequence be given equal weight. This principle has powerful intellectual appeal and intuitive force. To discredit a proposal or a theory it is often enough simply to show that it violates this principle. When measures are said to be discriminatory or unfair it is this principle which is in play.

Recent philosophers of widely differing orientations and beliefs have given this principle a central role in their theories. Ronald Dworkin, Robert Nozick, and Jonathan Glover for example have all given it a central place,[18] although it is fair to say that they would all disagree as to just how it is to be interpreted.

If people's lives and fundamental interests are of equal value then it is unjust to treat people differently in ways which effectively accord different values to their lives or fundamental interests. Deliberately to give one person a better chance of remaining healthy and have a long life than another is to value that life and that fundamental interest in health more than the person not so benefited. It is to discriminate in her favour. Where literally all cannot be benefited, equality requires that the method of selecting who will benefit and who will not is fair. This is why scarce resources which bear upon the value of life or the fundamental interests of persons must be allocated justly.

No dogs in mangers

One method of allocation of a scarce resource which apparently satisfies the requirements of justice is of course not to allocate that resource to anyone! All are then treated equally, in the sense that they are all left equally without benefit of the resource in question. This is often thought to be a viable application of the requirements of justice and indeed to constitute a just allocation of resources. The fallacy of such a supposition is easily illustrated. The principle of justice, and indeed the principle of equality, are *moral* principles. That is they are principles with some moral content, principles that are designed to be more than impartial, that are designed among other things to respect and to do justice to persons. In some sense this must involve some benevolent attitude to persons which is often abbreviated as 'respect for persons'. Such an attitude to others is as different as it is possible to be from simply showing *an equality of lack of respect* or *an equal indifference to the fate of others*.

The failure to allocate resources that would save lives or protect individuals could not then be part of a claim to satisfy the requirements of equality because this principle has at its heart the claim that people's lives and fundamental interests *are of value*. Anyone who denied life-saving resources, or resources which would protect life and other fundamental interests, is not valuing the lives of those to whom she denies these protections. Although she is treating them all equally in the sense of treating them all *the same*, she is not treating them as equals, as people who matter and hence matter equally.[19] The alternative dog-in-the-manger approach treats all people as *equally unimportant* and hence as equally without value.

In the case of biotechnological modifications to the human genome the problem of just allocation, of an allocation that treats people as equals, arises in a number of different ways.

Scarce resources

The first is of course the familiar problem of scarce resources. The technology necessary to achieve these genetic modifications is, for the foreseeable future, likely to be both scarce, or relatively scarce, and expensive. It will need highly trained staff, laboratory conditions, and so on. It will not be the sort of thing that could be done at a village clinic for example. These familiar constraints imposed by scarce resources create problems of just distribution which are difficult enough in themselves, but at least they are comfortingly familiar.[20]

Where resources are scarce two questions arise. The first is how urgent is it that people benefit from the deployment of these resources? The second question is how can we distribute these resources in a way which satisfies the principle of equality?

If the resources are life-saving, that is if they are required to prevent premature death, then it is clearly very important that they be deployed.

Hobbesian obligation

Arguably protecting citizens against threats to their very lives is the first priority for any state. This is the classic Hobbesian[21] argument for the obligation to obey the sovereign. Any citizen's obligation to the state and to obey its laws is conditional upon the state for its part protecting that citizen against threats to her life and liberty. In most contemporary societies the most significant of such threats come not from the threat of armed aggression from without, but from absence of health care and other social welfare measures within. This is why it is arguable that the obligation to provide health care, and in particular life-saving health care, takes precedence over the obligation to provide defence forces against external (and often mythical) enemies.

Whether the creation of new citizens who will be protected against many dangers and who are likely to live longer is on a par with, for example, rescuing accident victims who are in danger of their lives, is difficult to say. It is certainly not implausible to claim that it is equally important, and circumstances, like perhaps a failure to find any other method of stemming the spread of AIDS, might make it more important. In any event, it is clearly a highly important public health measure and is likely to have a high priority.

The answer to the first of our two questions then is that we should try to give as many citizens as possible the chance to have children who will be members of the new breed. This means giving it a very high priority in the allocation of public resources. The remaining question then is: given that not everyone who wishes to avail themselves of this resource will be able so to do, how should we determine its allocation?

The only morally respectable answer to this question is: in a way which does not violate the principle of equality. That is to say, in a way which does not unfairly discriminate against any particular individuals or groups of people.

Here of course it is easier to be clear about the moral principle which we should accept as governing our conduct than it is to know quite how to go about obeying such a principle. It is easier to show why certain established methods of allocation violate the principle than to arrive at one clear way of acting in accordance with it. However, any of the established ways of just distribution would be acceptable as a starting-point and we do not have to adjudicate between them.

The important thing is to accept a high priority for the distribution of the resources necessary to create the new breed. The distribution system then might be on a first come first served 'waiting list' basis with people postponing conception until the resources available to allocate to their children membership of the new breed had become available. Equally an annual lottery could be held for the allocation of places.

Circumspection

A second problem of scarcity arises because of the degree of circumspection required for any individual or couple to avail themselves of the benefits of this sort of genetic engineering even supposing it to be readily available.

If, as we have supposed, 'genetic protection' of a comprehensive sort is on offer in some sense (whether commercially or via the health service or in some other way) we must ask how would-be beneficiaries can avail themselves of such protection? Well, first of course they can only provide it for their children for, as we have noted, the protection has to be built in at the gametes or early embryo stage. For this reason also it will only be available to those who are sufficiently circumspect to seek it in advance of procreation. For candidates will either have to avail themselves of IVF techniques or will have to have their gametes modified. In either case only those prepared and able to plan ahead in this way will be able to benefit, and they will be able to benefit only those of their children who are 'planned'—for even planned families may have unplanned members.

This will again impose further problems for the just distribution of such a valuable benefit. For those capable and able to plan in this way will both be a minority, and are also likely to be a minority drawn from particular, and particularly advantaged socio-economic groups. These groups will then be further advantaged by having disproportionate numbers of their members benefiting from genetic protection.

Controlled benefit

We have so far refrained from preferring a method of allocation and the measures we have described are consistent with something like a free market constrained only by principles of just allocation, and with a customer-led service, in which genetic protection is in principle provided for those who seek it and scarcity is handled by some mechanism for the just allocation of a scarce resource. But of course another way of handling scarcity is for some sort of goal-directed control to be imposed from above, by a government say. We are I hope naturally suspicious of any such measures. One reason why a government might seek to impose such controls has already been rehearsed. It is of course the fact that there may be tasks a government would wish performed which could be best performed, that is performed most safely, by members of the new breed. How a government might secure the services of members of the new breed once they had reached the age of majority is a further question of course and one to which we will return.

Justice between societies

We have noted some of the ways that the various forms of scarcity may lead to injustice within a particular society capable of offering genetic protection to some of its citizens. When we consider the problem of justice between societies the problem of genetic protection serving the ends of justice becomes manifestly greater. In the first place we have already identified a preliminary problem when we mentioned 'societies capable of offering genetic protection'. For this will only, again for the foreseeable future, be industrialized, relatively wealthy societies, unless those societies are prepared to share their ability to provide genetic protection. But even if altruism rules, the scale of protection on offer in the non-industrialized world will inevitably be less than that available for 'home consumption'. Moreover, the problem of circumspection in the non-industrialized world will be proportionally greater than in 'the west'. There are a number of obvious reasons for this. In the first place, the knowledge of the availability of the techniques will be harder to disseminate, because there is likely to be a greater degree of suspicion of the techniques and perhaps more pervasive religious or cultural objections to interference with fertility and because the habits or the acceptability of birth control more generally will be likely to be less well established.

For these reasons and others, the introduction of genetic protection will inevitably widen the gulf between the industrialized world and the non-industrialized world.

Justice between individuals

The problem of justice between individuals would not of course be solved even if the problem of the just allocation of scarce resources could be solved. If for the moment we suppose we can solve the problem of who is to have access to genetic protection in a way which does not unfairly discriminate against particular individuals or groups or classes of individuals, or between particular societies, we will be embarking on a process that will inevitably sharpen people's perceptions of relative deprivation and injustice. For, where some people are protected against terrible infections like AIDS or hepatitis B and others are not, the sense that to contract a disease like AIDS is not simply a misfortune, albeit one of tragic proportions, but also an injustice, will be hard to escape.

And of course, particularly in the case of AIDS, the injustice will not only affect those upon whom the risk of catching AIDS crystallizes into a certainty, it will be an injustice to all who are *vulnerable* to AIDS. They will have to modify their lifestyles and take precautions that the protected individuals, the members of the new breed will not. These differences are likely to be substantial in a number of ways. Since sex is, not to put too fine a point on it, generally acknowledged to be one of the most important activities of life, the freedom to pursue one's sexual preferences without fear and without the necessity to be profoundly cautious or take elaborate precautions against the risk of infection will be a substantial benefit. Then of course if it is established that a particular individual has immunity there will be insurance and other financial advantages and perhaps even new job opportunities available.[22] In short the advantages accruing to the new breed will, as we have seen, be substantial and will exceed the simple (though not mere) immunity to disease and other disabilities which they will possess.

Of course there is a sense in which it is always an injustice when one person's health suffers relative to others, even where this is in some sense 'natural' and unavoidable. But because it is unavoidable, natural injustice is just a brute fact about the world, one that we may regret, but one of which it is bootless to complain. The injustice of being unprotected relative to others in society or indeed in the world, relative to the new breed in this case, is not a natural injustice. It is not like being disadvantaged relative to a Mozart or a Shakespeare. It is a disability which one might not, need not, have possessed. It is a disability that has been imposed by virtue of the advantages bestowed on others.

The next question that must be considered is the crucial one of when and in virtue of what does a relative disadvantage or disability become unjust or otherwise wrongful? To do so we must revive an earlier part of our discussion when we considered the case of wrongful life and examined the distinction between harming and wronging.

III. Harmful or Wrongful Disability

In a world in which the new breed exists and is becoming established, to lack the immunities and protections possessed by the new breed is to have a substantial relative disability. I hope this is self-evident but to those to whom it is not, consider the following scenario. A woman is a patient at an IVF clinic, she is told that the capacity to confer on her child substantial genetic protections exists and that if she wishes a few new genes can be added to her own embryo to confer on it the protections which will make her child a member of the new breed. The procedure is simple, carries no risks to the embryo if successful, and if unsuccessful she has spare embryos available to be substituted for this one. Should she accept the offer, would she be wrong to decline it?

A first point to note is that there are important differences between this and Parfit's example considered earlier.[23] With the exception of the capacity to repair DNA and so retard ageing, it is possible that the child will never be exposed to any of the dangers to which it might be rendered immune, and so will not in fact suffer from lack of the protections it might have had. We will assume that the DNA repair is not available to keep this example simple. And so unlike Parfit's child, it will not necessarily be injured by failure to carry out the genetic modifications. Now of course even this is not strictly true since, as we have seen, to be vulnerable though not actually wounded is still a disadvantage. There will be psychological dangers associated with this vulnerability as well as social and financial disabilities relative to the protected individuals. Moreover, knowledge of one's vulnerability will close options that might otherwise have been available.

On a trivial level it might be like saying that inability to play a musical instrument is not a disadvantage to those who never actually wish to play one. Or, more clearly it is like declining to protect say, an aircraft, against a possible (though perhaps statistically relatively remote) danger, with the consequent risk to every load of passengers it carries. The refusal to build in the protections is not less culpable because a particular aircraft never experiences the conditions which would trigger the disaster.

The more familiar obligation, which is more closely analogous, is the obligation to vaccinate young children against diseases which continue to pose a danger, albeit usually a statistically remote one for individual children.

This seems to me the right set of analogies and consequently the duty of care that parents have to their children extends to the duty to provide them with such genetic protections as are available, just as they have the duty to vaccinate babies against prevalent dangers such as polio and whooping cough where appropriate.

The failure to provide the available genetic protections would be to harm the child in two ways. To cause it actual harm by imposing on it the extra precautions and vulnerability plus the possible social and economic disadvantages that go with the vulnerability *and* to impose the increased risk of serious disease or damage relative to the protected fellow citizens.

Whether it would also constitute a wrong to that child would depend on whether or not that child might have been better off if the protections had not been applied.[24] This would be the case, where the next child of these parents could only be protected by aborting or failing to implant a particular fetus so that another protectable fetus could be implanted. A child who was not protected, and so exists whereas it otherwise would not, has not been wronged unless existence itself is a wrong to that child and, as we have seen, this will only be the case where existence is in sum a burden to that individual.

Where this is not the case, to fail to provide the protections would also constitute a wrong to that child.

The Winston Smith argument

There is a deeply ingrained rebelliousness in human beings which coupled with a love of spontaneity and freshness is always likely to undermine moral imperatives that require the sorts of radical circumspection which would enable people to avail themselves of genetic protections for their children. Winston Smith in George Orwell's *Nineteen Eighty-Four* is perhaps the apotheosis of this approach to life and it is difficult not to applaud its reckless humanity. Typical is this scene between Winston and Julia:

'Have you done this before?'

'Of course. Hundreds of times—well, scores of times anyway.' . . .

His heart leapt. Scores of times she had done it: he wished it had been hundreds—thousands. Anything that hinted at corruption always filled him with a wild hope. Who knew, perhaps the Party was rotten under the surface, its cult of strenuousness and self-denial was simply a sham concealing iniquity. If he could have infected the whole lot of them with leprosy or syphilis, how gladly he would have done so! Anything to rot, to weaken, to undermine![25]

A cult, if not of strenuousness and self-denial, at least of circumspection and self-denial would be required by would-be parents of the new breed. Even now there is some rebellion against the moral imperatives created by our knowledge of how habits, diet, in short lifestyle, affect the course of a pregnancy. Already people resent strictures to abstain from smoking, alcohol, drugs, and other things known adversely to affect the developing human individual. Inevitably, even if genetic modification of their children is available to them, many will see the circumspection required for such protection of their children to be not only burdensome but too great a sacrifice of other values, values like Winston's reckless humanity.

It is of course much easier (though not *that* easy) to sympathize with Winston Smith's wish to inflict leprosy and syphilis on self-righteous Stalinists than it would be to share the morality of those who wish to rebel against moral pressure to protect their children. But something of the same spirit is likely to be involved. A natural hostility to the enemies of freedom, spontaneity and 'natural' living.

The invocation of the spirit of Winston Smith seems simply misplaced here. Freedom and spontaneity are not unproblematic virtues and where they are substantially harmful to others we ought to be prepared to modify our behaviour and find other outlets for spontaneity.

The fear of AIDS (and before it the fear of pregnancy) has taken much of a certain dimension of spontaneity out of many people's sex-lives, but most people can see the force of the difference between spontaneity and recklessness. There is no doubt that recognition of our responsibility to others inhibits behaviour in all sorts of directions. The fact that we sometimes revel in throwing off our inhibitions does not, even temporarily, license rape or murder. We simply have to find other ways in which to be uninhibited.

For my own part I welcome the possibility of a new breed of

persons with life chances not available to us now. Of course engineering genetic protections of various kinds into the human genome will not automatically make the world a better place to live in, nor will it necessarily make people happier. We will still have to work as hard as ever to reduce disease, new diseases are after all always liable to arise. We will still have to work as hard as ever to reduce prejudice, including prejudice against the new breed, to combat injustice, to eliminate poverty, starvation, cruelty, and the thousand unnatural shocks that flesh is heir to; as well as the natural ones.

But the fact that we cannot cure everything has never been an argument for failing to cure something, particularly when it is something that causes pain, misery, and premature death.

10

Screening and Discriminating

There are screens and screens. When we screen a film, we make it visible, we put it on show. Sometimes screening individuals for particular characteristics, traits, conditions, or diseases is like this. It is a process whereby these traits, characteristics, conditions, or diseases are simply revealed and identified. But we also screen things in the sense of 'screening out' or filtering. And in this sense the screen is not a surface on which things are revealed but a filter through which some things are allowed to pass while others are caught and left behind.

Biotechnology has greatly improved our capacity to screen individuals in both senses. In the first sense it greatly increases the knowledge we have about people, both in general and in particular. This knowledge enables us to help those people of course and it enables them to help themselves. But it also greatly increases the sum total of reasons we have to discriminate concerning those people, both for them and against them.

We need to review the range of discriminations the new technology has made possible and then explore the ethical ramifications of these possibilities. We will start with a brisk canter over the ground to familiarize ourselves with the course and then look in more detail at those parts of the terrain likely to give us the most trouble.

The 'book of life'

As Mark Ferguson reports:

It is now possible to take a single cell, extract the DNA from it and then, if one knows the region of the genome one wishes to examine, to amplify up that region to obtain enough DNA for analysis by gene probing. Currently there are some diseases which are diagnosable using the quantity of DNA present in a single cell. This technology is rapidly expanding and it seems likely that nearly all future diagnoses could be made on the basis of a single cell biopsy.[1]

Almost certainly this single cell will be obtainable simply, either from the pre-implantation embryo growing *in vitro*, or, after birth, by a non-invasive procedure, probably using a mouthwash or mouth swab to collect saliva which would contain enough cells for the biopsy. This procedure would make possible the selection only of healthy embryos for implantation, or the genetic modification of embryos with disability or genetic damage prior to implantation. In adults it would provide a large volume of diagnostic material with immense potential, both for good and sometimes for things less than good.

When the human genome project is completed, or very soon thereafter, this process of single cell biopsy will enable the complete 'book of life' for each individual to be compiled and of course to be read. This will reveal not only the presence of genetic disorders which will or may affect the individual in question, but it will also make possible the comprehensive detection of asymptomatic carriers of defective genes. These carriers can then be given appropriate genetic counselling as to the risks to their children or other family members.

Equally the information contained in the book of life may yield techniques for the replacement or repair of defective genes *in situ*. As well as these dramatic therapeutic advantages the book of life will doubtless also reveal much trivia. It will, like most books dealing with lives, contain a great deal of information of minor importance, in this case concerning subtle but slight genetic differences between individuals. Some of these differences will indicate statistical risks for particular individuals which are insignificant both in themselves and when compared with the normal everyday risks of life. They might, for example, reveal that the individual is at 2% greater risk than average of contracting a condition that affects only 1 person in 100,000 in the population. Compared with the normal risks of, say, living in a congested city with its consequent pollution and traffic hazards this risk may be vanishingly small.

As well as providing a dramatically comprehensive method of screening individuals for disease and risk of disease, both to themselves and other family members, the mapped genome or book of life will reveal the likely susceptibility of an individual to various occupational illnesses and dangers. And not only these of course, there may also be revealed risks to individuals from particular living conditions, susceptibility to dust hazards for example. Other special risks which may be revealed can include those from wider environmental and climatic factors. These may range from special susceptibility to skin cancers from exposure to bright sunlight or to hazards associated with damp or humid climates to allergies associated with animals or plants.

We must also note the dramatic possibilities opened up for education by the new screening techniques. If genetic connections are established for things like musical ability, athleticism, and intelligence there will be immense pressure to specialize in the education of children earmarked for success or failure in such areas.[2]

All of this new information may well be double-edged. As well as providing information which may generate therapies and preventive strategies, it will also provide information that might be of interest to third parties, to other family members, colleagues, employers, insurance companies, and of course to the government or to government agencies. Most of these possibilities will be examined in more detail as the argument progresses but in the meantime we should backpedal slightly and look at the more routine ways in which these and other agencies have been interested in the more conventionally garnered information with which we are already familiar.

Screening for disease

We are familiar with many screening exercises in the context of disease or illness. Screening properly so called, is done at any routine visit to the doctor or the dentist or to a health care clinic or during any routine visit from a health care professional like a district nurse or health visitor. It may take the form of a 'good look' by the doctor or dentist or involve some questioning or some routine tests. These may range from taking blood pressure, testing sight, and so on, to cervical smears or urine or blood samples. We are more and more familiar with the ways in which such screening can have ramifications beyond our immediate or even long-term health care.

First and most importantly it generates information which is stored in medical and maybe in other records. This information may be sought by third parties of various kinds, by employers, by the government, by insurance companies, and even by spouses or other family members. The giving of such information may affect the individual's ability to obtain insurance cover, mortgages, and so on and also of course to obtain or maintain employment or a particular employment. And even if the information is withheld by the health professional concerned, this too can have important consequences for the patient. For the various agencies will draw inferences from the withholding of information and may act on these inferences to the disadvantage of the subject. Equally, agencies may make disclosure of information a condition of acceptance of the insurance proposal or the subject's candidature for employment or office or whatever.

Often under present circumstances a patient will not be aware of the contents of her medical record and so will not be aware, if she consents to disclosure, just what precisely might be revealed, nor what the effect of the disclosure will have on her chances of obtaining the desired service or employment. And, if she declines to permit disclosure or if the doctor declines to reveal information in the absence of the patients consent, she may not be aware what inferences will or may be drawn from this decision.

Pre-natal screening

The traditional problems concerning pre-natal screening are those of the diagnosis and treatment of disease in the womb and the question of the possible termination of pregnancy as a result of the screening process. These we examined at length in Chapter 3. Our present concerns with screening are of a different order.

Screening for desired characteristics

We started with familiar health screening but we should really have started further back with an even more familiar sort of screening. It is often simply called 'selection' and it is done all the time and perfectly respectably. We are selected by our partners, by our employers, by our fellow club members, and many others, on the basis of all sorts of obvious and less obvious characteristics we possess. Many attempts have been made to ensure fairness in such selection processes—at least as far as employment is concerned, even to the point of discriminating in favour of features that have been discriminated against in the past.[3] However, the sorts of screening that biotechnology is making possible create a new and to some extent unprecedented set of problems.

Genetic screening

As we saw in the last chapter there may be substantial employment advantages in particular types of genetic immunity or protection. The corollary of that is of course also true and we know that there are substantial disadvantages for those with genetic disease or who have a high risk of contracting genetic or indeed other diseases or who might be specially vulnerable to certain occupational risks. As Suzuki and Knudtson have noted:

> In some cases employers have consciously or unconsciously selected their employees on the basis of visible hereditary traits that they thought would render their workers less vulnerable to work-related hazards or illnesses. During the early part of this century for example, if you sought employment in a tar and creosote factory and happened to have been born with a fair or freckled complexion, your application might have been turned down simply because of your skin color. . . . Since the late nineteenth century it has been known that workers constantly exposed to tar creosote and certain other petroleum products in their jobs were more likely to develop skin cancer than others.
>
> In the competitive world of business, one can quickly translate an increased likelihood of contracting occupational disease into dollars and cents. Employee illnesses can mean missed work days, higher turnover of staff, increased training costs and, in recent years at least, spiralling expenses for job benefits and even disastrous lawsuits.[4]

If it was whites, and freckled whites at that, who were discriminated against in the creosote business, no one will be surprised to learn that blacks have had more than their fair share of this sort of discrimination. Sickle cell disease is a hereditary blood disorder found mainly in black populations. This feature of sickle cell disease was the spring for a piece of naked discrimination by the United States Air Force.

> Fearing that even a single copy of the sickle-cell gene might interfere with the oxygen carrying capacity of Afro-Americans exposed to high altitudes during pilot training, the Air Force Academy excluded them from flight school over a 10-year period, without adequate evidence to substantiate its concerns. In 1981, under pressure of a lawsuit and unable to demonstrate that low-oxygen conditions would precipitate a medical crisis in these heterozygous carriers, the academy reversed its discriminatory policy.[5]

These were early and half-baked attempts to use genetic predispositioning factors as a justification for discrimination in the workplace.

But with effective and comprehensive methods of genetic screening imminent the possibilities are much more serious.

We will now look in turn and in more detail at the possible ethical and legal consequences of comprehensive genetic screening of the type that will be available when the genome project is completed. Then, as we have seen, the full book of life may be compiled and read from a single cell taken by a simple mouthwash or even before implantation in the case of IVF-assisted reproduction.

In particular we will examine the problems associated with employment and occupational hazards to health, with insurance and its connection, via mortgages, with the ability to buy a home, and with the acute problems generated by prior knowledge of susceptibility to disease states. Finally we will consider the whole business of the ownership of knowledge and information and the rights of access to that information.

We should remember here and bear in mind that there is currently an international co-operative project to map the entire human genome and it may be completed (appropriately) before the end of the millennium. Mark Ferguson reminds us that

> the development of a genetic and physical map followed by complete sequencing of the estimated 3,500 million base pairs making up the human genome are now technically feasible: all that is required is the will and the money (about 50 pence per base pair). The time estimated to complete this massive task varies from 3 to 30 years depending upon the international resources devoted to the project; the ever increasing sophistication of DNA sequencing machines is likely to reduce the time and the cost in financial and manhour terms.[6]

Occupational screening

There are two very obvious ways in which the genetic constitution of individuals is relevant to their occupation. Their constitution can either make them more or less suitable for that particular situation. Having available the screening results for individual members of society will thus be a useful protective and selective tool. It will also carry certain dangers as we shall see. Another, perhaps less obvious use for occupational screening might be the monitoring of the genetic constitution of individuals throughout their lives. This would enable the early detection of damage to this constitution and the possible identification of the source of such damage.

1. Selection of suitable employees or 'genetic screening'

One obvious use of genetic screening in this context is to identify those individuals whose genetic constitution renders them more susceptible to occupational hazards of various sorts. The hazards to which some people might be genetically more susceptible than others might include noise and air pollution of various sorts, contact with chemicals or other dangerous substances at work, and nuclear radiation. This might arise as part of the job, as in the nuclear or defence industries, or, of course, where radioactive sources are standardly used in more everyday contexts as they are in health care, as sources of radiation for X-rays for example, but also for radiation therapy, sterilization processes, and the like. Equally there are an increasing number of food and other materials processing plants which use radiation as a sterilizing process. In some geographical locations there is naturally occurring radioactive gas. Radon for example occurs naturally in the granite of Cornwall and elsewhere in the United Kingdom and is a hazard to those living and working above it. As we know even sunlight may be a hazard in certain contexts, and where the ozone layer is thin above a certain location it may be highly hazardous to be exposed to direct sunlight in the heat of the day.

By identifying those at increased risk we can, in principle, both protect the individuals from running unnecessary risks and also of course protect their employers from absenteeism due to illness or from wasting training resources or other employee investment on 'bad risks'. Equally the employees, insurers, whether these are life insurance schemes, medical insurance plans or indeed national insurance plans of various sorts, will also be protected from preventable calls on their resources and the unnecessary or avoidable deployment of scarce health care resources will be saved.

So far so good. One immediate problem is of course that we are in most cases only talking about increased susceptibility to danger. This means that individuals will run an increased percentage risk of succumbing to one of the various dangers we have identified. A particular individual might not of course succumb at all, and for many the increased risk will be slight. Whether it would be justifiable to regard individuals so affected as in any sense unsuitable for a particular occupation or unsuited to a particular geographical location or way of life is highly questionable. The use of genetic screening for screening out employees who are potentially vulnerable then carries three substantial dangers itself.

Three dangers

The first is obviously that where employment opportunities are to some extent in short supply, individuals may be discriminated against in employment or denied access to a chosen career or forced to change their job because they are at slightly greater than average risk, a risk moreover which they themselves may be willing to run if fully acquainted with its nature and extent. There is a problem here of course as to whether their willingness to run such a risk means that they are also willing to be personally responsible for any costs accruing if they contract an occupational illness. The answer to such a question will of course in part depend upon whether the circumstances in which they succumbed to the hazard involved any negligence or other culpable act on the part of their employer or others.

The second danger is, as we noted in the previous chapter, that the ability to filter out 'at risk' employees may lessen the imperatives to clean up and make safe the workplace or the physical environment. Finally, workers identified as being 'at risk' will not unnaturally be anxious, perhaps for the rest of what may prove to be long and healthy lives, about their condition and its consequences.

2. Genetic monitoring

Instead of (or in addition to) filtering out vulnerable potential employees before they can be damaged by hazardous working conditions, genetic screening techniques could be used to control occupational hazards by monitoring the genetic constitution of workers at various points to check for damage to the DNA. Such tests might also be used to settle the question of whether or not particular workers had indeed suffered damage attributable to their employment. Such questions are notoriously difficult to settle at the moment and the availability of genetic tests for damage throughout a worker's career might settle these vexed questions and permit workers to recover appropriate compensation.

However, properly managed genetic monitoring could do much more than this. Starting perhaps in the embryo stage or at birth, but certainly prior to employment, and continuing at intervals throughout life, genetic monitoring might first enable very early detection of genetic damage to become widespread. It would then be possible to take immediate preventive and therapeutic measures when damage is detected and such screening would also facilitate rapid therapy. The preventive measures might take the form of alerting those managing the workplace or controlling the hazard to the fact that there was a danger which needed elimination or control and enable the rapid removal of the damaged worker and her fellow employees from the source of the damage. Equally, appropriate therapeutic measures, if available, could be quickly instituted and other workers could be monitored immediately for similar damage.

What should we do?

If genetic screening is to be used to determine the ability of specific individuals to work at particular occupations or in particular locations, then we need to think carefully about what particular level of risk or what particular dimension of risk justifies, for example, exclusion from employment. When we talk of the 'level of risk', we mean something like the specific percentage of increased risk that is associated with a particular condition. The dimension of the risk is the nature of the illness or damage that will result to the individual if the risk crystallizes on them, that is, if they do indeed succumb to the illness or condition to which they are identified as being susceptible.

There are a number of different dimensions to this problem and we should consider them separately.

Duties of employers

Among the constraints operating on employers are moral constraints. They ought not to discriminate unfairly in the employment opportunities they offer, neither should they put their employees at avoidable risk. We can say, therefore, that an employer would only be justified in excluding an employee or potential employee on the grounds of particular susceptibility to an occupational condition if a number of basic minimum conditions are fulfilled. We will first see if we can identify a plausible list of these conditions and then return to look at their precise meaning and implications. It may be of course, even with the safeguards we are about to examine, that ultimately we will reject all of them in favour of some quite different approach to the problem.

1. The employer should have done everything possible to eliminate the risk factor from the place of work. Such risks as remain must be effectively ineradicable and of the nature of the occupation in question and must be more generally ethically acceptable.
2. The employer must make the employee aware, in advance of employment, of any remaining risks or make the employee aware of them as soon as they become apparent. That is to say the employee should give informed consent to any occupational hazards.
3. The presence of any risk factor in the employee must represent a risk in cost or some other palpable risk, either to the employer or to others who share the working environment significantly above the normal level of such risks inherent in a free and mobile work-force operating in the market-place.
4. Such risk as the presence of the employee represents to the employer or third parties must be sufficiently severe as to make it both unreasonable to expect an employer to bear such a risk or to permit third parties to run that risk, and so severe as to justify the consequent discrimination against the employee.

At first sight, it would seem that the first two conditions should be part of the duties of any employer regardless of that employer's possible role in excluding candidates for employment on the grounds of their supposed susceptibility to such hazards as remain. However, it might be argued that it is unfair and inefficient to force employers to clean up the workplace at high cost and force customers to pay more, when genetic screening might produce a work-force for whom the risks were acceptable. This raises a very important question:

What is a safe workplace?

If we go back to the first condition we can see that while simply stated it is very complex in detail and would be very difficult to encapsulate in legislation. It is obvious, I hope, that it would be both dysfunctional in terms of public health and safety, and also inequitable, to allow genetic screening to enable employers to renege on their obligation to provide a safe working environment: one moreover which is safe not only for those who share that environment, whether employees or not, and also safe for those who live within range of the possible dangers it represents.

If we use the nuclear industry as an exemplar of an industry presenting one such possible set of occupational hazards we can see what is involved here.

Chernobyl has shown the ambit of the danger represented by nuclear power to be wide indeed (many thousands of miles wide) and the combination of Chernobyl, Three Mile Island, and Sellafield have shown that such dangers as are represented by nuclear power are not so exceptional as to be effectively non-existent.[7] Equally there is much evidence now tending to suggest the more localized dangers of, for example, increased incidence of leukaemia near nuclear power stations.[8]

Now if there is to be nuclear power at all there are clearly ineradicable dangers in the nature of the industry, processes, and materials involved. These ineradicable dangers affect both those working in the plant and the outside world. It follows that unless there is an overwhelming public interest in the maintenance of nuclear power, all such installations are unethical more generally, in the sense of representing a danger not justified by the countervailing advantages of the availability of nuclear power. Moreover, a danger to which those exposed have not consented. And the relevant public interest of course is not that of a single society merely, but of all those within range of the possible harmful effects.

It seems incontrovertible that we should not permit a workplace to become a major public danger. However, many occupational hazards, machine noise, dust, physical contact with noxious chemicals, and so on, present dangers which are largely confined or confinable to the workforce. Here there seems to be some initial attractiveness in a tailor-made workforce, but there are problems.

The first problem is of course that increased resistance is not the same as complete resistance. The question of how much danger it is reasonable to ask a workforce to face, particularly when those dangers could be reduced or eliminated by the institution of safety measures, is one that genetic screening does not remove.

Secondly we would have, as a matter of public policy, to face the question of whether we should encourage the practice of genetic discrimination in the workplace. For not only will there be pressure or inducements to persuade workers with enhanced protections to take jobs that would otherwise be even more dangerous, but there is likely to be pressure between employers, or agreements (or conspiracies)[9] of employers which will result in the managers of safe or safer workplaces declining to take on employees who might be more 'advantageously' employed elsewhere. The power that this would give employers to manipulate the workforce is, or could be, immense.

While working conditions are unsafe and improvable it would seem to be overwhelmingly in the public interest to ensure that such workplaces are made safe, both because the enhanced immunities and protections revealed by genetic screening or achieved by genetic engineering are unlikely to be total, and because screening like all other diagnostic tools is not immune to the possibility of false positives. But perhaps chiefly because pollution of all sorts is rarely completely confinable to the workplace and we are all safer if such hazards are eliminated.

However, there are cases where the public interest will be served by encouraging the employment of individuals with genetic immunities.

To take a less controversial example, as long as sources of radiation are necessary in the provision of health care, whether as sources of X-rays or as therapies or as sterilizing agents, then a small radiation danger will be ineradicable in the sense I mean, from the occupation of health care provision. Of course all steps available must be taken to ensure employee and public safety as far as this is possible. But, since there is an overwhelming public interest in the provision of health care generally and of its having available radiation therapy, and sterile equipment in particular, and moreover, since those working in health care are aware of and accept the small risks involved, the first condition is in this particular case met. How then will the third and fourth conditions operate? We will take them together.

The right to work?

These latter two conditions address the question of whether or not the refusal to employ someone who will be particularly susceptible to occupational hazards is unjust, or perhaps violates that employee's rights. This is a complicated question for there are many different possible rights and interests involved here.[10] There is on the one hand society's legitimate interest in public health and in preventing citizens being unnecessarily exposed to health hazards, particularly when these hazards cost public money to treat and use up scarce resources. Then there is the fairness or otherwise of imposing on employers increased and unreasonable costs, by preventing them from selecting employees who are suitable in the sense of being cost-effective. Next comes the issue of protecting third parties from possible danger represented by diseased or unreliable workers. Finally, there is society's equally important interest in protecting the rights of citizens and in particular protecting them from unfair discrimination. How are we to balance all these interests?

I have said this is a complicated question, and indeed it is. But complex issues are not necessarily difficult to resolve and this one seems more straightforward than most. There seem to be few arguments for diminishing the employer's responsibility to provide a safe working environment both for employees and for all others who may be affected by that environment, however remote geographically or temporally. And such arguments as there are apply only to cases, such as those we have reviewed, where there is a particular and strong public interest in the utility of a particular workplace and its safety cannot be further enhanced except by the employment of a more hazard-resistant workforce.[11] Although the employer might face some increased costs from operating a non-discriminatory employment policy, no one has ever claimed that respecting rights or doing what is right costs nothing. Indeed the whole point of the idea that people matter morally is that their rights or fundamental interests cannot simply be cast aside when it is convenient or cost-effective so to do.

It might be argued that this approach is question begging. If a particular employer needs 100 employees and there are 500 applicants, why should she not choose the 100 who are most hazard resistant rather than, say, spend an extra million pounds in cleaning up the workplace? In particular, whose rights are violated if employees are regarded as specially qualified by their pollution resistance status and hence chosen in preference to others?

We have already argued that there are independent reasons to require the employer to clean up the workplace even at substantial cost, reasons which do not speak to the issue of employee rights or the obligation not to discriminate against less resistant employees. These reasons will normally be sufficient. However, there does seem to be a residual argument against discrimination here.

If we think it wrong for employers to be permitted to discriminate against women on the grounds that they are statistically likely to wish to take time out to have children and that they are consequently less cost effective as employees than men, we do so because we recognise that the right to work or not to be discriminated against in the provision of employment opportunities is important and that the added costs of such a policy are such as should be borne in order not to violate these rights or in order to do justice. Equally, people should not be ruled out for employment opportunities for analogous reasons connected with their genetic constitution and the added costs that possession of such a constitution might impose on the community.[12]

Discrimination against particular individuals, or particular types of individuals, in the provision of employment is a major wrong, whether that wrong is analysed in terms of denial of respect for persons, or in terms of a violation of their rights.

Ronald Dworkin puts this point most eloquently and his conclusions bear repeating:

> So if rights make sense at all, then the invasion of a relatively important right must be a very serious matter. It means treating a man as less than a man, or as less worthy of concern than other men. The institution of rights rests on the conviction that this is a grave injustice, and that it is worth paying the incremental cost in social policy or efficiency that is necessary to prevent it. But then it must be wrong to say that inflating rights is as serious as invading them. If the Government errs on the side of the individual, then it simply pays a little more in social efficiency than it has to pay; it pays a little more, that is, of the same coin that it has already decided must be spent. But if it errs against the individual it inflicts an insult upon him that, on its own reckoning, it is worth a great deal of that coin to avoid.[13]

This argument of Dworkin's seems to me powerful and happily (for me) consistent with the consequentialist approach taken in this book. I do not believe that rights have any special status, that is, I believe that rights always emerge as the conclusion of a moral argument as to what should be done rather than, as in Dworkin's argument, as one of the principal premisses in any such argument. So that when we have decided for example that certain individuals should not be discriminated against in employment, we may express this as protecting their rights. However, Dworkin's argument here is consistent with such an approach, for the argument is simply that discrimination attacks the obligation to treat people as equals, and accord to each the same concern respect and protection as is accorded to any. Any policy which has the consequence of violating the equality principle is worse than one which merely has some increased monetary or efficiency costs, costs which, as Dworkin points out, we have already committed ourselves as a society to bear. This is because the equality principle is at the centre of our social morality, whether consequentialist, or not, and is presupposed by democratic theory.

Dworkin's theory of rights rests on the equality principle for he states 'The institution of rights rests on the conviction that [treating people other than as equals] . . . is a grave injustice'. It also seems to me[14] that the equality principle must be at the heart of consequentialism, a rather different moral approach to the one favoured by Dworkin. The idea that persons matter morally is the idea that each matters morally. If each person matters morally, then it matters what happens to each person, and hence we must, when making decisions, consider what the consequences will be for anyone affected.[15] If this were not so then there would be no point in considering consequences at all.

Of course there are versions of consequentialism that apply the equality principle not to persons, but to units of happiness or to preferences summed independently of the persons whose preferences they are, or to years of lifetime. However, the argument of this book has been that it is persons that matter.[16]

If, further, we defend the right of individuals to take and keep jobs even at some risk to themselves we presuppose of course that the individuals are aware both of their genetic profile and of the possible danger represented by a particular occupational environment. If they then freely choose that particular occupation it would be wrong to deny them employment opportunities on account of their very nature —their genetic constitution. I believe this to be so even where, in exceptional circumstances, the risks to a particular individual are very great. She should then be counselled against hazardous employment, but not denied such employment if she is otherwise qualified or suitable for it.

There is of course a danger of exploitation here, and we have already reviewed in some detail the circumstances in which exploitation, or the danger of it, is itself a moral objection to permitting a practice. However, with one exception to which we are about to come, the danger of, and motives for, exploitation in this area are much less than the dangers of unfair discrimination. It is seldom in an employer's interests to employ people who are particularly susceptible to occupational hazards for reasons we have reviewed. However, it may well be in the employees' interests not to be barred from a job they desire to do, because it might be somewhat more risky for them to do it. In the past we have seen campaigns for the disabled not to be discriminated against in the workplace even though they may be somewhat more at risk than other employees.[17]

However, in order for those with genetic susceptibility to danger to be protected, we must ensure that they are not driven by desperate need to seek hazardous employment.

It has been suggested that there is substantial social utility in permitting those with raised susceptibility to occupational hazards to be permitted free employment opportunities because, rather 'than penalising them for their perceived defects, we might look on them as warning beacons representing the final boundary beyond which society should no longer condone further environmental contamination of a workplace'.[18] It looks rather as though this suggestion treats people as latter day pit canaries, which were kept with miners underground to detect lethal gas emissions. When the canaries died the miners knew it was time to leave!

The lesson from this is that we should not permit contamination of a workplace at all and should certainly not use 'human guinea pigs' to detect it. However, there are circumstances in which we will have to accept some risk of contamination and in which some people may be more susceptible to damage if this risk crystallizes on them. In such a case the decision to run the risk should be left to the fully informed individual.

To take just one example, if I have some genetic susceptibility above the norm to radiation damage, I should not be barred from working as a health professional because I will very likely have frequent resort to a workplace in which there are sources of radiation.

Conclusion

Genetic monitoring and genetic screening, like so many things, can be used for good or for ill. We should not forgo the benefits for fear that we will not have the courage to outlaw the harms. If instituted, both will require back-up in the form of counselling services and treatment where appropriate. Equally we will need legislation to control the potential for discrimination that these tests will release. With these provisos genetic screening and monitoring will increase human autonomy and help us to protect both individual people and the environment.

11

Record Breaking

As well as enabling us to monitor the health of individuals and to identify their genetic constitution, the new technology is a compulsive generator of information. The screening and monitoring opportunities which we have just reviewed will create an ever increasing body of data on the individual. Much of this will of necessity have to be stored on computers and the problems of access, privacy, confidentiality, and maybe even of ownership of the information will become acute.

There will also of course remain the already familiar problems associated with genetic screening. Among these are the problems that may arise because information about one family member of necessity yields information about genetically related individuals. An intelligent patient will know, or be able to work out, that the information she is being given about her condition could only have arisen because one of her siblings, or parents, is known to have a particular genetic constitution. Equally, there are the agonizing difficulties associated with having to tell someone that they carry, or will contract, an incurable and perhaps devastating condition.

I. Insurance

We will begin by looking at the business of risk assessment and of insurance against risk, for, as will be obvious, the ability to carry out screening and monitoring to an unprecedented extent will have a profound affect on the insurance market and on the things which the existence of such a market currently makes possible.

Assurancetourix

The title of this section is of course also the name of a famous bard, the companion of the celebrated 'Gaulois' and cartoon character Astérix.[1] Assurancetourix might be rendered less sonorously into English as Allrisksinsurance. Such insurance is doubtless very useful, even essential in a modern society, and perhaps the most vital part of the insurance market is the life insurance industry, closely followed by medical and health insurance. Most people require a mortgage in order to become home owners and virtually all mortgages require some form of life insurance cover. In the United Kingdom, where the government of Mrs Thatcher deliberately starved the National Health Service, medical and health insurance has become ever more important, and of course in societies like the United States, there is little alternative but to secure private health insurance.

As screening and monitoring become more comprehensive and more specific and as the predictive capacity of the information generated becomes more accurate, the population will begin to divide into two significant groups—the uninsurable and the free riders. We should be clear from the outset that it will not be possible to keep the new and ever increasing data generated by genetic screening out of the hands of insurance companies for the simple and sufficient reason that they can demand it as a condition of contract. If it is withheld, insurers will simply deny insurance cover, and of course false or incomplete information will vitiate the policy and automatically render the candidate uninsured.

The uninsurable

These will not of course be a homogenous group, nor will most of them be literally uninsurable, though they will be progressively hard to insure. Those diagnosed as having conditions which will cause premature death or substantial disability will be veritably[2] uninsurable. They will, like most HIV+ people today, be unable to obtain life insurance and mortgage cover and will find substantial difficulty in gaining employment. Others will find insurance cover very difficult to obtain or obtainable only on payment of unrealistic premiums.

Good risks

Some people will be very good risks, they will have the sort of profile that indicates resistance to many genetic disorders or other conditions with a genetic component. They may even be members of the New Breed if breeding it ever becomes a reality. In all events they will present minimum risks to insurers. Of course they will remain susceptible to the proverbial 'Number 43 bus', but many standard risks will have disappeared. These good risks will qualify for their reduced premiums in a sense, at the expense of those excluded from insurance altogether or those with heavily 'loaded' premiums. Of course since they pay premiums too they will not be 'free riders' and since total premiums payable by those allowed to subscribe will presumably be less because some of the highest risks have been excluded altogether, they do not benefit from their reduced premiums at the expense of those with heavier premiums. They do, however, ride at a tremendous cost to those not allowed to participate and for them the costs are highest (though not of course in terms of raised premiums).

Risky business

The existence of these two groups will of course make life much easier for insurance companies. The *risk* which hitherto has been of the nature of insurance business will progressively be reduced. We have already witnessed a substantial shift from *assessment of risk* as the basis of insurance to *minimization of risk*. This is the process whereby insurance companies, instead of accepting that the name of the game is to assess the risk of a particular policy and set premiums accordingly, have gradually moved to attempts to minimise or even eradicate, rather than simply assess, risk. This process will inevitably accelerate with the ideal being the dismissal of risk altogether, insurance becoming a form of regular saving.

If (or perhaps when) this process is completed, we will have moved from *insurance against all risks* to *insurance only of the risk free*.

National insurance

There seems to me to be only two ways of resolving this problem, and one or other of them must be mandatory. It is clearly grossly inequitable to deny people the benefits of insurance because they are prudent enough to be screened and monitored for health risks. It may seem somewhat naïve to say this since of course this gross inequity is already standard practice. In the eyes of insurance companies, you are among the best risks if you are so imprudent as never to visit the doctor. The more you visit or are visited, and in particular the more investigations you have, the worse risk you are deemed to be. If, and you had better not, you are imprudent enough to be investigated for something *specific*, if you are foolish enough for example to undergo investigations for heart disease, cancer, or AIDS, then, whatever the result of these investigations, you will become a 'worse risk' for insurance purposes.

The most notorious and iniquitous current example is of course testing for HIV+ status. If you have such a test then even negative results will automatically put you in the 'high-risk' group. One of the effects of this is that conscientious doctors advise many patients not to be tested at all or to be tested at an anonymous clinic or (in the United Kingdom) by the blood transfusion service—where neither the results nor the fact of testing will show up on the medical record. This practice, forced on caring doctors primarily by the insurance industry, has the effect of decreasing the chances of early detection for many people, with the consequent increased risk to their sexual partners and others who may be in touch with bodily fluids. This state of affairs is disastrous both from a public health perspective and also from the perspective of personal relationships. People are faced with the alternative of ruining their lives by placing themselves and their family in an almost impossible insurance position and in an increasingly precarious employment position or literally risking the lives of their families by ignoring what may be a slight but real risk.[3]

Of course some, maybe many, who would wish to be tested wish this because they *are* in a high-risk group. However, many may simply be anxious or over-anxious. There is after all a first time for even a monogamous relationship and all such relationships could in principle turn out to have been 'one night stands'. Equally people may have come into chance and isolated contact with blood or other bodily fluids in circumstances which do not change their general risk status and simply want reassurance.

Both we, the citizens, and successive governments have so far been rather spineless in allowing this gross injustice to continue. The massive increase in screening will be a suitable occasion to reassess our attitudes to such injustice.

I mentioned that there are two ways in which these inequities might be removed. They are these: the first would be simply to abolish all life and health insurance from the private sector and create a national insurance system which would provide cover for all, irrespective of prognosis or diagnosis. Premiums would have to be equal for all citizens and could either be set at a level to enable this to be achieved, or if, as may be likely, this would make premiums too high for some, then these could be subsidized by the taxation system. The second strategy would be to impose the obligation to provide insurance for all, with equalized premiums, on the private sector as a condition for being permitted to stay in business.

Now, perhaps I have gone too far for two reasons. The first is that the schemes, as briefly outlined, would seem to be unworkable. I have not dealt with obvious problems; like what to do about the terminally ill patient who seeks, for the first time, comprehensive and massive life insurance. The second reason is perhaps that remedies as specific as these are beyond a reasonable remit for a philosopher. However, we can say something about them.

Realism

I think that the practical problems are usually overplayed. Premiums should be set as if there were no information available, as if screening were not a feature of existence. If they were set for average life expectancy or illness, or accident expectancy for a particular age, the risk would even out. We must remember that for every terminally ill 20-year-old who might get 'unfair' cover there would be many whom screening and monitoring would protect from disease or premature death and who would consequently *balance* those who cash in.

Since screening is likely to reduce the vulnerability of the community at large, insurance companies cannot be worse off than they are at present if they are required to set premiums as if no new information existed. For this reason legislation constraining insurance companies might be less controversial than would at first appear.

II. Control of Information

We have noted how for insurance purposes it is virtually impossible to control the information generated by screening and monitoring. It may be demanded as a condition of contract and inferences may be drawn from refusal of permission to provide information or to question medical advisers. The same problems will arise in the context of employment and of release of information to government and government agencies.

Another problem with information about genetic constitution and health status or risk status is simply the difficulties of keeping information secure. Computers, for example, are notoriously leaky and, in the context of health care, so are many 'professionals', from doctors and lab technicians to receptionists and cleaners or porters, who may have access to information.

In the context of genetic counselling, as we have seen, it may be possible for one family member to draw inferences about the genetic constitution and risk status of other family members simply from what she legitimately knows about herself.

Then there is the whole vexed problem of confidentiality. We have noted some of the brute difficulties in the way of keeping information confidential even where there is a clear willingness and felt obligation so to do. A real question of course is how far this obligation stretches in the face of powerful moral imperatives for breaking confidentiality.[4]

Finally there is the complicating factor of conventions, including of course, legal conventions, about ownership. In a property conscious society, a society which often prides itself on the degree to which its social and political institutions are market lead or ape the market-place in other ways, the issue of the ownership of information is always likely to be tested in the courts.

All these factors make the control of the information which is being generated and will increasingly be generated by the new technology a real and pressing problem. If we look at the problem under each of the heads we have considered we may find our way through to a solution of the problem of just how we might legitimately control such information, or perhaps even whether or not we should try to do so. We will start with the issue of ownership.

Property

What sort of property is information? A first and obvious answer is of course that it is 'intellectual property'. This is a form of property well known to the law and its more familiar dimensions have to do with copyright law or the law of patent. At the moment it is more than doubtful that information about a person's state of health, or even about his genetic constitution could count as the sort of intellectual property that might appropriately be patented or be the subject of copyright. For one thing you cannot patent or copyright the physical description of yourself nor indeed facts about your state of health. So long as I am truthful I am entitled to give as complete a description of you as is available to me and as can sustain the interest of another person. Of course I should not deliberately misdescribe you, and if I do give false information about you which would 'tend to lower you in the estimation of right thinking people generally' my action may be actionable, I may be guilty of slander or libel. But if I am truthful *and* my words are true, then I will have a complete defence to the charges of libel or slander.[5] This will be true in jurisdictions such as the United Kingdom which do not recognise a 'right to privacy'. Where, as in some states in the United States of America, such a right is recognised the question will turn on the extent to which my description violates this right.

Now I may, of course, have come by the information in ways that render me vulnerable to legal redress or I may be breaching a confidential relationship in publishing it. These are separate matters to which we will return.

I may have dealt too smartly with the question of property in personal information. There are two recent cases in the United States which are of some interest in themselves and which bear on this question. Both concern the question of whether or not a person 'owns' his own body or parts of it and also owns the information yielded by the study and use of bodily products.

In the first case a Japanese researcher, Hideaki Hagiwara, claimed property in a cell line derived from his mother's cells, for which therapeutic potential was claimed. He eventually 'stole' or borrowed back the cell line and used it to treat his mother. The dispute between his family and the University of California at San Diego was eventually settled out of court.[6] The other major cell line case has only recently been resolved by the Supreme Court. It raises more clearly the central issues.

John Moore v. *The Regents of the University of California*

In September 1984, John Moore, a patient who had been successfully treated for 'hairy cell leukaemia' (HCL), a very rare form of cancer, sued the University of California 'on the grounds that two UCLA researchers took unfair advantage of him by using his cells as a basis of research that has lead to a patent of undetermined financial value'.[7] The facts seem to be these: Moore's cancer treatment involved the removal of his spleen, for which operation he signed a consent form relinquishing his spleen and apparently consenting to its being used for research. Research on cells removed from the spleen carried out by David Golde and Shirley Quan, resulted in the development of a productive cell line which they called 'Mo' after Mr Moore, on which a patent was finally filed in 1981. This cell line produces various substances of value to scientific research, including immune interferon (type II), macrophage-activating factor, and T-cell growth factor. Mo is for example the cell line from which Dr Robert Gallo of the National Cancer Institute of the United States isolated human T-cell lymphotropic virus (HTLV-II).

As Sandra Blakeslee reports in *Nature*, 'The potential commercial value of the Mo cell line lies in its unusual lymphokine activity. Lymphokines are hormone-like substances secreted by lymphocytes with wide-ranging biological activity. They act locally to modulate cells responsive to them.' These substances have many potential uses and at the time the patent was first applied for there was particular interest in their possible role in the production of interferon as an antiviral agent.[8]

Physical property

Much of the legal argument surrounds the issue of whether or not Moore was entitled to the patent or a share in it simply in virtue of the fact that it was a patent obtained in part at least on his own tissue. Now, patents are not usually thought to be applicable to natural objects nor to properties inherent in things existing in nature. Rather they are obtainable only on things an inventor does with those properties. If this is right,[9] then Moore's claim seems problematic in the extreme.

However, Moore has several other possible causes of legal remedy. The two most promising are 'conversion' and the issue of the validity of the consent obtained. Conversion is a civil wrong in which one person 'converts' to his own use the property of another without necessarily laying hands on it or removing it. It may also take place where property legitimately in someone's hands for one purpose is wrongfully 'converted' by them to another use. In cases like that of Moore, conversion would probably have taken place if blood or tissue consensually supplied for one purpose was 'converted' to another use, one which *ex hypothesi* would not have been consented to by Moore under those conditions had he known about it at the time. This is particularly likely to be the case where the property is converted for profit.

If, as is clearly the case, Moore's informed consent was a prerequisite of the University of California validly obtaining the right not only to remove tissue and blood from Moore but to use it for further purposes, in this case research and development of a commercial product, then certain conclusions can be drawn. If the facts are as reported, then in, as it were, inviting Moore to contribute blood and tissue under one set of assumptions and then using his bodily products for another and self-enriching purpose, the University of California would certainly have wronged Mr Moore and vitiated the validity of his consent. However, this simply means that they would have been guilty of an assault and a battery on Mr Moore and would have in all probability breached their own code of practice. More important for the question of financial redress is the chance that they would also have violated the equitable doctrine of unjust enrichment.[10] This principle of legal equity recognises the injustice of someone benefiting from a wrong done to others, in short the wrong of exploitation.

Intellectual property

So far we have considered the issues on the assumption that it was the misappropriation of physical property, albeit physical property of a controversial sort—bodily products—that was at stake. However, part of Moore's case may be that the University of California also violated his rights to intellectual property or information. As Barbara Culliton suggests:

> Although the patent on the Mo cell line clearly states that the cells were derived from Moore's spleen, his attorneys hope to show that blood drawn from their client on several occasions in the years he was under treatment after the splenectomy contributed significantly to the research that lead to the patented cell line.[11]

The claim here is of course mixed. The lawyers will want to argue that the physical property, the blood in this case, was wrongfully used to obtain the knowledge required for the development of the cell line and hence that this too was a case of conversion. But an important part of their case here is likely to be the claim that the blood taken was not required for therapy or routine follow-up monitoring but was taken deliberately and deceitfully so that it could be used in potentially profitable research.

There may be a question here as to whether or not Moore gave permission for the use of his blood or spleen in research and as to precisely what his permission encompassed. Did it for example extend to profit-making which excluded Moore himself or merely to altruistic research? The crucial question seems to be not: does the alleged wrongful act involve conversion or theft of *property*? But rather does the alleged act constitute an abuse of information or of the providers of the information?

Suppose no physical property was wrongfully taken or converted. Suppose that by examining a patient with the naked eye, a procedure for which consent was not required, a researcher gained information which she could use to develop a commercially viable product. This might I suppose occur where a patient has a relatively rare or even unique condition, important clues to the curing of which were visible without any invasive procedure for which consent was required. Should the patient here have any claim on the use made of this information or the ideas which flow from it?

Or suppose that a researcher scans computer records of data on a number of patients, records which were legitimately acquired and responsibly stored. Suppose she discovers something important, and perhaps also commercially viable, from this? Two questions immediately arise. The first is: is her initial decision to scan records for scientific or therapeutic research legitimate? The second is: is her commercial exploitation of the results legitimate? These are complicated questions.

Before addressing them, however, we should note that the law of property is currently the only remedy available to those who feel themselves exploited, hence the pressure to ground alleged wrongful exploitation firmly in the theft or conversion of *property*, and the understandable pressure to think of information in terms of property; just as in the case of 'wrongful life' litigation which we examined in Chapter 4, where although inappropriate, the only avenue open to disabled children requiring financial help with disability and the huge expenses it causes was to take legal action against their parents for compensation for the alleged 'wrong' of bringing them into existence in a disabled state. We should, of course, be careful not to deprive individuals of their only avenue of redress, however inapposite it may be, without replacing this avenue with something better and more sensible.

Public interest

There is no doubt that scientific, medical, and therapeutic research is in the public interest. We want people to learn from experience and we want medical and scientific knowledge to grow. We want this not only because we ourselves will benefit; the benefits may be more remote and may not in fact advantage us at all. Research is justified not only for selfish reasons but because others and future generations will also benefit. A researcher who believes she might learn something from our records would be negligent not to consult them if she had the time and the opportunity and could legitimately do so.

The provisos we have already entered here beg many of the most important questions. What do we mean by stipulating that the researcher has the time, the opportunity, and might legitimately consult our records? Of course people might well reasonably believe they could learn something of use or importance from our records and still not have time or opportunity for all sorts of reasons: because they had more important things to do, because this research was not funded or could not obtain funding, because it was already being undertaken by someone else, because the right to consult the records belonged to someone else, etc., etc. This last point also of course speaks to the issue of legitimacy. It might also fail to be legitimate to consult my records for any number of other reasons: because they were confidential and the particular researcher was outside the circle of confidentiality.[12]

Within practical constraints, it is in the public interest for the persons most likely to make productive use of the records to have access to them. By practical restraints I mean simply that it might not be possible for more than a few researchers to have access to a particular set of records at any one time. Or it might be unfair to allow free access to records which were being productively used by one group; this might simply permit rivals to 'steal a march' on them.

The main questions we must consider here are whether the public interest is served by allowing the individual about whom the records are information to possess a veto on their use or to possess the rights to dispose of them to whomsoever he wishes. In the absence of commercial considerations and in the presence of strong conventions about confidentiality and even stronger protections for the individual, I believe the answer to both these questions to be a firm 'no'.

Of course, the public interest is not the only interest here, but it stands for the interests of countless present and future individuals who will benefit from research and the open access to information for research purposes. The individual's right to privacy or to control of information about herself is not obviously as powerful or as important a consideration as that of preventing disease and premature death in the population at large. This does not of course mean that the individual is completely vulnerable to those with a plausible research protocol nor that she is fair game for exploitation. The safeguards must be such as to permit the individual who is the source of the information to share in any commercial rewards which flow from it and to be protected from breaches of confidentiality which might adversely affect her.

Just what is meant by 'sharing' in commercial rewards here we will come to in a moment. We must of course also remember the point made above, that natural objects are not usually patentable although of course they may be capable of being property.

Genetic endowments

There is also a difficulty which I must record but cannot here discuss. It is this: in so far as information which is usable for research and is commercially viable is genetic information, and in so far as such information may be owned, it is not clear that the ownership does not stretch back to Adam and Eve and beyond to their primate progenitors! This point is discussed at length by Hillel Steiner in his forthcoming book *An Essay on Rights*[13] and I leave its ramifications in his capable hands.

Commercial exploitation

There are two ways in which the individual might share in any rewards resulting from the commercial exploitation of information of which she is the source. One is by giving her the equivalent of copyright in the information, so that she will share *pro rata* in the proceeds of any profits made from this information. The second is by allocating to the public purse the share in any profits due to the providers of information.

This latter course might be more sensible or the more defensible for three reasons. The first is that calculating the proportion of profits due to an individual, information about whom has made an infinitesimally small contribution to a commercial product, might be horrendously difficult. For example, where information from hundreds or even thousands of individuals itself comprises a minor proportion of the contribution to a final product and where the information from each might have made a different size contribution, the cost of calculating the share might well be more than the share itself. There would also be the added problem and cost of record keeping over long periods of research and development. In such cases the better, the more efficient and cost-effective course might be simply to determine the proportion due to the experimental subjects or information donors and allocate their share of the profits to the community at large.

Finally there is the problem of where or rather with whom or with what the information originates. If Hillel Steiner is right to suppose that genetic information stretches back to our primate progenitors of human beings and if animals are plausibly classified as 'natural resources', then since many theories of justice see natural resources as public property, this gives an added plausibility to the claim that profits should be part of the public purse. However and of course there is a further problem about how public this purse should be. For in so far as common ancestors are ancestors of humankind as a whole it would be difficult for particular nation states to claim that their small partisan corner of humanity should share the profits exclusively.[14]

Since the public purse is a large vessel and its contents are apt to be spent on all sorts of frivolous purposes like nuclear deterrence and indeed the nuclear industry generally, and since the information we are talking about will usually have been derived from patients or in the context of health care, we should stipulate that these profits go into that sector or the public purse devoted to health care and disease prevention.

In some circumstances it might be unjust to do this. Where researchers or their employers are making large profits from information (or bodily products for that matter) and where it is not simply dysfunctional or self-defeating to allocate profits to individual contributors, it might be unfair to permit researchers or institutions to make the profits and prevent these profits reaching those without whom there would have been no profits.

In such circumstances it might be economical to leave such matters to be settled contractually between donor and researcher. Of course safeguards would have to be built in to prevent researchers exploiting the vulnerability of donors who may also be patients.[15]

A problem here of course is it will probably be difficult to know in advance which scenario is likely to prove operative in a particular research project; and the problem of deciding whether or not to keep records and set up a mechanism for tracing contributors at the outset might kill off many promising initiatives. For this reason I think the argument comes down in favour of legislation allocating profits from the donor side of research to public health.

This of course still leaves the problem of deciding how to determine the relative contribution made to a product by information gained from research subjects or patient records on the one hand, and the expertise of researchers on the other. This is the sort of problem that philosophers are happy to leave to economists and other technicians.

The instability of 'other things'

The second problem presented by the existence of increasing volumes of personal information is that of security and confidentiality. Other things being equal people are entitled to expect that information given in confidence, or information arising from tests consented to on condition that the results are confidential, will remain confidential. The problem is that 'other things' tend to be radically unstable, and though they may start off equal, they are likely to be destabilized by all sorts of unforseen considerations. This is not the place to rehearse the familiar problem of confidentiality,[16] but one example will illustrate the point. People tested for AIDS, for instance, may demand confidentiality, but on the one hand they are not entitled to place others in risk of their lives just so that their confidentiality be preserved. So that where their infection with the HIV places others at real and present risk of infection, those others are entitled to know so that they may protect themselves. On the other hand, as we have seen, for insurance purposes and the like, it is unlikely that individuals can avoid the consequences of non-disclosure, for insurance companies will draw inferences from a refusal to answer questions.

Either way, the protection of confidentiality is not one that can be relied on. There is a third reason for this. Much confidential information may have a commercial value, whether, for example, to insurance underwriters or to the popular press. Information about the health status of those in the public eye and the further clues this yields as to the private lives of the famous or notorious, is always in demand. Where this is the case, determined investigators are likely to achieve access to information, however carefully stored and whatever conventions or rules about privacy or confidentiality prevail. The solution must be to protect individuals not from the release of information but from at least some of the worst consequences of that release.

Protection and the right to privacy

In the case of insurance we saw that one way of doing this would be by a national insurance scheme which refused to notice such leaked or revealed information and set its premiums accordingly. The only other effective way to protect individuals from the damaging consequences of release of information is by legal penalties and financial compensation set at levels which will be sufficient to deter.

To do this effectively we would need to consider some formal recognition of a right to privacy. The only effective way this could be done is through legislation. Doubtless this would be immensely complex and its effect on such issues as freedom of speech and of the press would have to be carefully thought through. However, with the explosion of information already under way such consideration is long overdue.

III. Knowledge is a Dangerous Thing?

I do not of course suppose or believe that all this new information will be an unequivocal blessing. There is no doubt that we will discover all sorts of things about ourselves and others that we might have preferred not to know. Many of these things will worry us unnecessarily. Other things will of their nature be worrying and we would be crazy not to be worried. While it is a truism that 'we should not be unduly worried', and this platitude is constantly repeated by people offering spurious assurance, the appropriate question is always: *how duly worried should we be?* And we can never answer such a question unless we have the appropriate information.

Most of us are carriers of something

When the genome is mapped and genetic screening is fairly routine, we are bound, for example, to discover that we carry various anomalous genes which are not damaging to ourselves but possess the ability to pass on a genetic disorder to our offspring if we mate with another carrier. Current screening techniques are already revealing facts like this to countless individuals. Almost all of us carry some of these anomalous genes and at the moment most of us will never know that we do. For most abnormalities the chance of meeting and mating with another carrier and passing on a defect will be rare. There is some evidence that where individuals are aware of facts like this they are disinclined to wish for screening of potential partners and even where they may have reason to suspect a potential partner is a carrier are unlikely to decline partnership on this ground. A recent report states that no 'study reported to date has shown that genetic counselling alters marriage patterns'.[17] I do not know whether these facts are comforting. However, at least as many as are unduly worried will be duly worried, and will be pleased to have the opportunity of genetic screening of partners and of pre-natal screening and genetic counselling where appropriate.

Ethnic susceptibilities

Various ethnic groups have particular susceptibilities to genetic disorders and members may consequently wish to benefit from screening. Tay-Sachs disease particularly affects Ashkenazi Jewish populations and, as the report of the Royal College of Physicians on Pre-natal Diagnosis reminds us, was 'the first recessively-inherited disorder to be prevented at the community level by carrier screening and the offer of pre-natal diagnosis'.[18] Communities with a strong tradition of consanguineous marriage are also particularly susceptible to some disorders, for consanguinity increases the chances that two carriers of the same recessively inherited condition will mate. Among those who are vulnerable to this risk are British citizens of Pakistani origin where the rate of first-cousin marriage is 55% as opposed to 32% in Pakistan. Here again, population screening could be advantageous.[19]

Equally, some of us will discover that we are more susceptible to particular diseases than others because of our genetic constitution. Here again the amount of due worry this should occasion depends upon the degree of the increased susceptibility. In many cases it will be small, a few chances in tens of thousands or whatever. However, the degree of undue worry such knowledge may occasion should be more than offset by the chance such forewarning gives of decreasing other risk factors for a particular condition that may be within the agent's control. For example, someone aware of an increased genetic susceptibility to heart disease will have the opportunity, which they might otherwise have neglected, of attempting to control other risk factors for heart disease, like smoking, cholesterol consumption, obesity, and lack of exercise.

Use and abuse

All techniques can be abused and there is no knowledge or information that is not susceptible to manipulation for an evil purpose. The task is always to be vigilant and to prevent such abuse. We do not abjure food for fear we will become obese, although we may become obese if we are not careful.

New procedures always provoke dire warnings and yet the advances in information gathering and dissemination made possible by biotechnology are not radical departures from the familiar fears about the relation between knowledge and power. Spectres have of course been raised of a world in which screening is compulsory and only the genetically compatible and 'healthy' are permitted to breed, or one in which only the genetically 'pure' were marriageable and in which certificates of genetic hygiene had to be presented to potential partners. If we fear such a world we must make sure it does not become a reality. But our distaste for such possibilities should not prevent us from availing ourselves of the benefits of genetic screening. As always and as with any innovation whatsoever, it is up to us to decline the burdens.

Notes

Introduction

1. The *Oxford English Dictionary*, 2nd edition, gives two definitions of bio-technology: 'The branch of technology concerned with the development and exploitation of machines in relation to the various needs of human beings'; 'The branch of technology concerned with modern forms of industrial production utilizing living organisms, esp. micro-organisms, and their biological processes'. The *International Dictionary of Medicine and Biology*, vol. i (John Wiley & Sons, New York, 1986) suggests: 'The application of the biological sciences, especially genetics, to technologic or industrial uses'. Clearly the *International Dictionary of Medicine and Biology* is closest to what I believe is the central idea but the sense in which I use the term is consistent with both sources.
2. Although I am aware that biologists have a rather more restrictive use for this term, thinking of it principally perhaps in connection with industrial uses.
3. Shakespeare, *Julius Caesar*, Act II, Scene i.
4. This I have done elsewhere. See my *Violence & Responsibility* (Routledge & Kegan Paul, London, 1980) and my *The Value of Life* (Routledge & Kegan Paul, London, 1985).

Chapter 1

1. D. J. Weatherall, 'The New Genetics', in Jonathan M. Austyn (ed.), *New Prospects for Medicine* (OUP, Oxford, 1988) 43. This essay is an excellent introduction to the new genetics. Other good introductions for the general reader are: E. Yoxen, *The Gene Business* (Pan, London, 1983) and T. A. Brown, *Gene Cloning* (Van Nostrand Reinhold, Wokingham, 1986). See also Stephen G. Oliver and John M. Ward, *A Dictionary of Genetic Engineering* (CUP, Cambridge, 1985).
2. Weatherall, 'The New Genetics', 47.
3. Mark W. J. Ferguson, 'Contemporary & Future Possibilities for Human Embryonic Manipulation', in Anthony Dyson and John Harris (eds.), *Experiments on Embryos* (Routledge, London, 1990). See also Professor Ferguson's 'Dentistry and the New Biology', Darwin Lecture for the British Association for the Advancement of Science, 26 Aug. 1987,

forthcoming in the *British Dental Journal*. It will I hope be obvious that I am greatly indebted to my colleague Mark Ferguson both here and elsewhere in this book.

4. Ferguson, 'Contemporary & Future Possibilities', 6.
5. Ibid.
6. See my 'In Vitro Fertilization: The Ethical Issues', *Philosophical Quarterly*, 33: 132 (July 1983).
7. Ferguson, 'Contemporary & Future Possibilities', 10.
8. R. Edwards and J. Purdy (eds.), *Human Conception In Vitro* (Academic Press, London, 1981), 373.
9. T. G. I. Creteil, 1 Aug. 1984. *Gaz. Pal.* 1 (1984), 560. See brief discussion in D. Giesen, *International Medical Malpractice Law* (Kluwer, Paris, 1988), para. 1359.
10. Ferguson, 'Contemporary & Future Possibilities', 6.
11. Ibid. 11.
12. See Carol Ezzell, 'First Ever Animal Patent issues in United States', in *Nature*, 332: 21 (1987). See also Christopher Joyce, 'Patent Law Protects Altered Organisms', in *New Society* (30 Apr. 1987).
13. Ferguson, 'Contemporary & Future Possibilities', 13.
14. Ibid.
15. See Chs. 5–9.
16. In chs. 5–7.
17. Ferguson, 'Contemporary and Future Possibilities', 15.
18. Weatherall, 'The New Genetics', 60.
19. Ibid. 68. For details of the possibilities of wound healing see Ferguson, 'Dentistry and the New Biology'.
20. R. G. Edwards in a personal communication.
21. See John Harris, *The Value of Life* (Routledge & Kegan Paul, London, 1985 and 1989), 166 ff.
22. A. J. Jeffreys, V. Wilson, S. L. Thein, 'Hypervariable "Minisatellite" Regions in Human DNA', in *Nature*, 314 (1985), 67–73.
23. The doubts about DNA fingerprinting arise largely from the 'Frye Hearing' procedure in the United States, which governs the acceptability of new scientific evidence, the presumption being against such evidence. See *Frye* v. *United States*, No. 3968 Court of Appeals of District of Columbia 293 F. 1013 1923. For an authoritative commentary on this procedure see J. E. Starrs, 'A Still-Life Watercolour: Frye v. United States', *Journal of Forensic Sciences*, 27: 3 (July 1982), 684–94. For commentary on the effectiveness of DNA fingerprinting see Thompson and Ford, 'Is DNA Fingerprinting Ready for the Courts', *New Scientist* (31 Mar. 1990), and Pamela Knight 'Biosleuthing with DNA Identification', *Biotechnology* (June 1990).

24. Some at least of these questions are discussed in Chs. 10–11 and in Chs. 5–6.
25. Edwards and Purdy, *Human Conception In Vitro*, 380.
26. Ibid.
27. Ibid. 381.
28. Ibid. 382.
29. Ibid.
30. Ibid.
31. For example work progressing at Dulwich and reported in the *Guardian* (1 Aug. 1988).
32. Conventionally biologists regard the genotype as consisting of nuclear genes and these are of course, by definition, all in the nucleus.
33. I mean 'incalculable' literally to mean 'not susceptible of calculation', not, as it is commonly used, as a synonym for 'huge'.
34. See Ch. 2.
35. Robert Edwards, 'Ethics and Embryology: The Case for Experimentation', in Dyson and Harris, *Experiments on Embryos*, 49, 50.
36. See Ch. 5.
37. See e.g. *Hoener* v. *Bertinato*, 67 NJ Super 517: 171A 2d 140 (27 Apr. 1961) for a United States case in which a mother purported to refuse a blood transfusion which would benefit her unborn child. A mother who refused a therapeutically indicated Caesarean in the United Kingdom was the case of *In Re R.* (*In Utero*) The Times Law Report, 5 Feb. 1988.
38. Reported in the *Guardian* and indeed in all major United Kingdom newspapers, 1 Aug. 1988.
39. J. Rifkin, *Confessions of a Heretic* (Routledge & Kegan Paul, London, 1985), 42.
40. See Ch. 5.
41. See Rifkin, *Confessions of a Heretic*, 48.

Chapter 2

1. In the first part of this chapter I will be deploying arguments already rehearsed in my 'Embryos & Hedgehogs', first published in the book I co-edited with Anthony Dyson, *Experiments on Embryos* (Routledge, London, 1990). This chapter, however, substantially revises the text of that essay.
2. Robyn Rowland, 'Making Women Visible in the Embryo Experimentation Debate', *Bioethics*, 1: 2 (Apr. 1987), 179.
3. Lady Warnock prefers the title 'chairman'.
4. Mary Warnock, 'Do Human Cells Have Rights?', *Bioethics*, 1: 1 (Jan. 1987).

5. Ibid. 1.
6. John Harris, 'In Vitro Fertilisation: The Ethical Issues', *Philosophical Quarterly*, 33 (1983), 217–37.
7. Isaiah Berlin, *The Hedgehog & the Fox* (Weidenfeld, London, 1967).
8. We will be returning to this issue, but for an eloquent account of this see Richard Dawkins, *The Selfish Gene* (Oxford University Press, Oxford, 1976).
9. H. W. Jones, jnr., 'The Ethics of In Vitro Fertilization—1981', in R. Edwards and J. Purdy (eds.), *Human Conception In Vitro* (Academic Press, London, 1981).
10. I have argued in more detail about the moral status of the embryo in my *The Value of Life* (Routledge & Kegan Paul, London, 1985 and 1989), ch. 1.
11. Society for the Protection of the Unborn Child leaflet entitled 'Back the Alton Bill', Jan. 1988.
12. For brevity I shall often use the term 'embryo' and the term 'fetus' as equivalents. Though there are some circumstances in which it is relevant to distinguish between them for our present purposes this will not be necessary.
13. Mary Warnock, 'In Vitro Fertilization: The Ethical Issues II', *Philosophical Quarterly*, 33 (1983), 238.
14. Warnock, 'Do Human Cells Have Rights?' 10.
15. See e.g. John Marshall, 'The Case against Experimentation', in Dyson and Harris, *Experiments on Embryos*.
16. Martin Johnson, 'The Onset of Human Identity and its Relationship to Legislation Concerning Research on Human Embryos', *British Journal of Family Planning* (1988). See also D. G. Whittington, 'Parthenogenesis in Mammals', *Oxford Reviews in Reproductive Biology*, 2 (1980) 205–31.
17. See e.g. a number of the reports in Marilyn Monk and Azim Surani (eds.), *Genomic Imprinting* (The Company of Biologists Ltd., Cambridge, 1990).
18. Susan Kimber has explained these points to me.
19. I am assuming here the moral equivalence of acts and omissions. I have argued for this equivalence at length elsewhere. In my *Violence & Responsibility* (Routledge & Kegan Paul, London, 1980) and *The Value of Life*, ch. 2.
20. *The Report of the Committee of Inquiry into Human Fertilization and Embryology* (HMSO, London, July 1984)—the Warnock Report, s. 11. 22, p. 66.
21. Warnock, 'Do Human Cells Have Rights?', 8.
22. Ibid.
23. Human Fertilization and Embryology Act 1990, ch. 37, clause 37, (HMSO, London).

24. Ibid. clause 3.
25. David Hume in his *A Treatise of Human Nature* (1738). Contemporary philosophers with a similar approach include Stuart Hampshire, see e.g. his *Morality & Pessimism—The Leslie Stephen Lecture* (Cambridge University Press, Cambridge, 1972); and Bernard Williams in e.g. his 'Against Utilitarianism', in B. Williams and J. J. C. Smart, *Utilitarianism For and Against* (Cambridge University Press, Cambridge, 1973).
26. Warnock 'Do Human Cells Have Rights?', 8.
27. Mary Warnock, *A Question of Life* (Basil Blackwell, Oxford, 1984), p. xv.
28. Warnock 'Do Human Cells Have Rights?', 8, 9.
29. Patrick Devlin, *The Enforcement of Morals* (OUP, London, 1965). I am indebted here to Ronald Dworkin's critique of Devlin in his *Taking Rights Seriously* (Duckworth, London, 1977), ch. 10.
30. In another context I called this approach to ethics 'olfactory moral philosophy' in deference to some remarks of George Orwell on the importance of using one's nose when thinking about right and wrong. See *Violence & Responsibility*, ch. 7.
31. Ronald Dworkin shows one way of doing this. See Dworkin, *Taking Rights Seriously*.
32. My argument here is derived from Ronald Dworkin. See ibid. ch. 10. My colleague Graham Bird has helped me greatly to clarify my ideas in these sections and I am grateful to him for a detailed critique of an earlier version of this chapter.
33. See Edwards and Purdy, 373.
34. Ibid. 381.
35. In his essay 'Ethics & Embryology: The Case for Experimentation', in Dyson and Harris, *Experiments on Embryos*, *Human Conception In Vitro*, 50.
36. This was anticipated in 1981. See Edwards and Purdy, 373 ff. The use of embryonic tissue is important because whereas human brain cells do not regenerate in adults they do in embryos and if embryo cells are transplanted into adults these cells may continue to grow and may stimulate adult cells to regenerate also.
37. Edwards, 'Ethics and Embryology', 49 ff.
38. See e.g. Robert Edwards, 'Causes of Early Embryonic Loss in Human Pregnancy', in *Human Reproduction*, 1 (1988), 185–98. See also H. C. Liu, H. W. Jones and Z. Rosenwackz, 'Efficiency of Human Reproduction after *In Vitro* Fertilization and Embryo Transfer', in *Fertility and Sterility*, 49 (1988), 649.
39. See my *Violence & Responsibility*, 48–65.
40. See Harris, *The Value of Life*, ch. 1.
41. Even if of course this 'chance' is a slim one because of the tendency of

anomalous embryos to abort spontaneously and of course some anomolous embryos are so anomolous as to have no chance of survival to a live birth let alone beyond.

42. See Harris, *The Value of Life*, ch. 1.
43. Edwards, 'Ethics and Embryology'.

Chapter 3

1. Robert Nozick, *Anarchy, State and Utopia* (Blackwell, Oxford, 1974), 337.
2. See my *The Value of Life* (Routledge & Kegan Paul, London, 1985 and 1989) and also for example Sue Mendus and Martin Bell (eds.), *Philosophy and Medical Welfare* (Cambridge University Press, Cambridge, 1989). Particularly the contributions of Anne Fagot-Largeault, G. R. Dunstan, and M. H. Johnson.
3. The discussion of the ramifications of this problem continues when we discuss wrongful life in the next chapter.
4. *In Re B.* (1987) 2 All Eng. 207 and also *T.* v. *T.* (1988) 1 All Eng. 613.
5. John Stuart Mill, *On Liberty*, ch. 5. in Mary Warnock (ed.), *Utilitarianism* (Fontana, London, 1972).
6. Derek Parfit, 'Rights Interests and Possible People', in S. Gorovitz (ed.), *Moral Problems in Medicine* (Prentice-Hall, Englewood Cliffs, NJ, 1976).
7. All that I have to say here derives from Derek Parfit. See his *Reasons & Persons* (Oxford University Press, Oxford, 1984), 366 ff. and 489 ff.
8. Ibid. 489.
9. Quoted by Parfit, ibid.
10. Thomas Nagel, 'Death', in his *Mortal Questions* (Cambridge University Press, Cambridge, 1970), 7.
11. See Ibid. 8.
12. See e.g. Michael Tooley, *Abortion and Infanticide* (Oxford University Press, Oxford, 1983), and my *The Value of Life*, for accounts of related conceptions of the person.
13. Parfit, *Reasons & Persons*, 490.
14. Ibid. 489.
15. Ibid. 490.
16. A technique called micro-injection involves isolating a single sperm and injecting it under the zona pellucida and is used to treat some forms of infertility. I am grateful to Peter Singer for reminding me of this possibility.
17. For my attempt to do this see n. 2 above.
18. See n. 2 above and also Peter Singer, *Practical Ethics* (Oxford University Press, Oxford, 1976). The law also draws a similar distinction in certain circumstances.

19. Parfit, *Reasons & Persons*, 489.
20. Ibid.
21. Ibid. 490.
22. This Parfit accepts when discussing for example population policy (ibid.).
23. See S. A. Kripke, *Naming and Necessity* (Harvard University Press, Cambridge, Mass., 1980).
24. Perhaps I have been slightly hard on the term 'Zygotic Principle' since, unlike the normal employment of the idea that life begins at conception, it does not imply that morally important life begins then. However, if we take the expression 'life begins at conception' literally it exactly expresses the Zygotic Principle.
25. Bernard Williams, 'Who Might I Have Been', in CIBA Foundation Symposium 149, *Human Genetic Information: Science, Law and Ethics* (John Wiley, Chichester, 1990), 169. See also Bernard Williams, *Problems of the Self* (Cambridge University Press, Cambridge, 1973).
26. Or where at gastrulation, the point at which the so-called 'primitive streak' forms, there form in fact two primitive streaks resulting in two identical embryos. See Anne McLaren, 'Prelude to Embryogenesis', in CIBA Foundation Study Group, *Human Embryo Research: Yes or No?* (Tavistock, London, 1986).
27. Williams, 'Who Might I Have Been', 175.
28. Williams had indeed discussed monozygotic twinning in his paper but had appeared not to see the full force of the problem it posed for ZP.
29. Ibid. 175.
30. If there are two gametes. The theoretical possibility of parthenogenesis makes it important to take seriously the idea that there may be (and may have been!) individuals who derive from only one gamete.
31. We shall discuss these possibilities again at length in Ch. 6–8.
32. See Chs. 7 and 10.
33. See Parfit, *Reasons & Persons*, esp. chs. 10–15.
34. Ibid. 262.
35. Ibid. 302. See also David Wiggins, 'Locke, Butler and the Stream of Consciousness', in A. Rorty (ed.), *The Identities of Persons* (University of California Press, Berkeley, 1976).
36. Personal identity on the other hand is quite another matter, it is about the identity of full blown persons and has to do with things like 'psychological connectedness', 'continuity of mind', etc.
37. Ibid. 271.
38. See *The Value of Life*, ch. 1.
39. I say 'appeared' because the second and third women may have failed in their attempts.
40. I deliberately do not specify the degree of handicap.

41. I am ignoring the various people other than the fetus or potential person that the mother might wrong in having an abortion. For example the father of the fetus who wants to have a child, or the members of SPUC (the Society for the Protection of the Unborn Child) who may be outraged, etc.
42. See Steven Hirsch and John Harris (eds.), *Consent and the Incompetent Patient* (The Royal College of Psychiatrists, London, 1988).
43. In a paper presented to the MA Health Care Ethics Course, Centre For Social Ethics and Policy, University of Manchester, 1987.
44. Claims about resource allocation I have discussed, *inter alia*, in my 'QALYfying The Value of Life', *Journal of Medical Ethics*, 13: 3 (Sept. 1987).
45. I do not of course pretend that this solves the problem of cruelty to particular children, because it may not of course be the parents who are responsible for the harm and it may be extremely difficult to find out just who is responsible.

Chapter 4

I am indebted to Len Doyal for his very helpful and perceptive comments on an earlier version of this chapter.

1. In Ch. 3 and of course John Stuart Mill, *On Liberty*, ch. 5, in Mary Warnock (ed.), *Utilitarianism* (Fontana, London, 1972).
2. Mill, *On Liberty*, ch. 5.
3. See *Zepeda* v. *Zepeda* (1963) 41 Ill. App. 2d 240. 190 NE 2d. 849.
4. *Gleitman* v. *Cosgrove* (1967) 296 NY S (2d) 689.
5. *Curlender* v. *Bio Science Laboratories* 165. Cal. Reptr 47 (1980).
6. See Bonnie Steinbock, 'The Logical Case for Wrongful Life', in *The Hastings Centre Report* (Apr. 1986), 15–20.
7. Ibid. 16.
8. See n. 4 above.
9. *Mackay* v. *Essex Area Health Authority* [1982] 2 All ER 771.
10. See Margaret Brazier, *Medicine Patients and the Law* (Penguin, Harmondsworth, 1987), 172. I am indebted here as so often to Margaret Brazier for her generous and scholarly advice.
11. Ibid. 167.
12. Ibid. 172.
13. See J. K. Mason and R. A. McCall Smith, *Law and Medical Ethics* (Butterworths, London, 1987), 101.
14. Joel Feinberg, *Harm to Others* (Oxford University Press, New York, 1984).
15. Above, n. 1.

16. Steinbock, 'The Logical Case for Wrongful Life', 19.
17. I wrongly, as I now think, attributed a slightly more extreme position to Feinberg in the earlier version of this argument. See my 'The Wrong of Wrongful Life', *Journal of Law and Society*, 17: 1 (Spring 1990).
18. In his *Harm to Others*. See also Derek Parfit's major treatment of these themes in his *Reasons & Persons* (Oxford University Press, Oxford, 1984), esp. chs. 16 and 17. We have discussed some of Parifit's views already in the previous chapter.
19. Feinberg, *Harm to Others*, 102.
20. Ibid. 103.
21. Ibid. 103, 104. Feinberg continues his theme in the latest and last volume of his study, *Harmless Wrong-doing* (Oxford University Press, New York, 1990), esp. ch. 31.
22. Ibid. 102.
23. I have argued this point at length elsewhere. See my 'Wrongful Birth', in D. Bromham, M. E. Dalton, and J. Jackson (eds.), *Philosophical Issues in Reproductive Medicine* (Manchester University Press, Manchester, 1990) and in Ch. 3 of this book.
24. Ibid.
25. Feinberg, *Harm to Others*, 99.
26. See the argument of the previous chapter.
27. Feinberg, *Harm to Others*, 99.
28. I ignore the pedantic issue of whether or not, strictly speaking, death can be a harm because death is not an event experienced by the dying person.
29. Feinberg, *Harm to Others*, 98.
20. Ibid. 102.
31. Steinbock, 'The Logical Case for Wrongful Life', 19.
32. Peter Singer convinced me of this latter point.

Chapter 5

1. See Reports in British national press for 5 Apr. 1990, esp. the *Guardian*, 1, 4.
2. This issue I have examined elsewhere: see *The Value of Life* (Routledge & Kegan Paul, 1985, 1989), ch. 7, and 'Surrogacy', in William Walters (ed.), *Human Reproduction: Current and Future Ethical Issues* (Balliere's Clinical Obstetrics and Gynaecology, in press).
3. In *Monty Python's Flying Circus*, the notorious BBC television series in which a parrot was the subject of the remark.
4. Regrettably, there are of course some people who appear to believe that the protection of their sensibilities is something for which others should

die, as the case of Salman Rushdie, the novelist 'condemned' to death by the late Ayatollah Khomeini illustrates.

5. I am here as elsewhere assuming the moral and causal symmetry of acts and omissions, assuming in short that decisions with the same consequences have the same moral status, whether they are decisions to do things or not to do things. I have argued for this position at length elsewhere. See my *Violence & Responsibility* (Routledge & Kegan Paul, London, 1980).
6. See e.g. David Lamb, *Death, Brain Death & Ethics* (Routledge & Kegan Paul, London, 1987). See also his *Ethics and Organ Transplants* (Routledge, London, 1990).
7. For an extended discussion of the problems of defining death see my *The Value of Life*, ch. 12.
8. The importance of the fact that these individuals *will never be* persons we considered in Chs. 2–4.
9. See Ch. 1 for a more detailed account of the possibilities here and also Mark Ferguson, 'Contemporary & Future Possibilities for Human Embryonic Manipulation', in Anthony Dyson and John Harris (eds.), *Experiments on Embryos* (Routledge, London, 1990).
10. Ferguson, 'Contemporary & Future Possibilities', 22.
11. Ibid. 23.
12. In Ch. 1.
13. See Robert Edwards, 'Ethics & Embryology', in Dyson and Harris, *Experiments on Embryos*.
14. Ibid. 49.
15. Ibid.
16. I have examined the question in different terms in my 'The Right to Found a Family', in Geoffrey Scarre (ed.), *Children, Parents & Politics* (Cambridge University Press, Cambridge, 1989).
17. Well, it is *some* evidence, assuming my sight or its intellectual equivalent is reasonable.
18. The biologist Richard Dawkins has most effectively identified and ridiculed a version of the argument from myopia which he has called 'the argument from personal incredulity'. See Richard Dawkins, *The Blind Watchmaker* (Penguin, Harmondsworth, 1986), 38–9. Like most authors I naturally prefer my own terminology, which I first used in 1976.
19. See e.g. *Medical Monitor* (26 Jan. 1988), 1.
20. See Chs. 2–3 and the more detailed argument in my *The Value of Life*, ch. 1.
21. 22 Apr. 1988.
22. The other arguments she produces we have considered elsewhere in this book, although we have used other writers as representatives of these positions.

23. See e.g. Arthur L. Caplan, 'Should Foetuses or Infants Be Utilized as Organ Donors?', in *Bioethics*, 1: 2 (Apr. 1987), 135.
24. The case history I describe happened in the United Kingdom. A similar one was recently reported from the United States in which Mary Ayala chose to have a second child as a bone marrow donor for her 17-year-old daughter Anissa. Reported in the *Sunday Times* (1 Apr. 1990).
25. See also Harris, *The Value of Life*, ch. 1.
26. I have also discussed this question at length in my 'The Right to Found a Family'.
27. The test of whether or not it is beneficial to use one child as a donor for another is one that has been a cornerstone of legal opinion on this question. See e.g. Margaret Brazier, *Medicine, Patients and the Law* (Penguin, Harmondsworth, 1987), 275, 276, for a critique of some of the assumptions implicit in this reasoning. See also the discussion of *Strunk* v. *Strunk* below.
28. Generally accepted to be about 1 death in 10,000 general anaesthetics, although the risk varies as to context—mortality being more likely where general anaesthetics are given without the major back-up available in a large hospital.
29. See e.g. William H. Bay and Lee A. Herbert, 'The Living Donor in Kidney Transplantation', in *Annals of Internal Medicine* (1987; 106 C5), 719–27, and A. Spital, 'Life Insurance for Kidney Donors—An Update', in *Transplantation* 4514 (1988), 810–11. In this study it was reported that in a sample of American life-insurance companies, all would insure a transplant donor who was otherwise healthy and only 6% of companies would load the premium. I am indebted to Søren Holm for pointing these sources out to me.
30. Unlike the cases of embryo and fetus donors we have so far considered.
31. As a result of the Gillick judgement we cannot assume even in law that these will necessarily be all minors. See *Gillick* v. *West Norfolk and Wisbech Area Health Authority* [1986] AC 112, [1985] 3 All ER 402, HL and My 'The Political Status of Children', in Keith Graham (ed.), *Contemporary Political Philosophy* (Cambridge University Press, Cambridge, 1982).
32. *Strunk* v. *Strunk* (1969) 35 ALR (3d) 683.
33. See also *Little* v. *Little* (1979) 576 SW (2d) 493. I am indebted to Margot Brazier for pointing these two United States cases out to me.
34. Though we must remember the standard risk from general anaesthetic.
35. This is not of course a perfect example for our purposes because the fate of the sick daughter if she was captured by the Nazis would probably be rather worse than death. However, we can assume that the mother either did not realize this or that in this particular case she would simply be immediately shot.

36. Of course these are not the same but for present purposes that is all I mean by 'wicked'.
37. See my discussion of 'The Survival Lottery', in *Violence & Responsibility*, ch. 5.

Chapter 6

1. See for example the discussion of intellectual property in Chs. 9 and 10. Also Jeremy Waldron, *The Right to Private Property* (Oxford University Press, Oxford, 1988), J. O. Grunebaum, *Private Ownership* (Routledge & Kegan Paul, London, 1987).
2. Or perhaps in exchange for an early appointment for sterilization, or a free one in a private clinic. This practice has been condemned by the *Independent Licensing Authority* in the United Kingdom in its 5th Annual Report published in 1990.
3. Perhaps simultaneously but in different ways, one sexually the other emotionally, for example. It may be of course in these circumstances that there is some disparity of value even here. One person valuing the other at rather less than par.
4. Joel Feinberg, *Harmless Wrongdoing* (Oxford University Press, Oxford, 1990), 199.
5. *Report of the Committee of Inquiry into Human Fertilization and Embryology*, Chairman Dame Mary Warnock DBE, (HMSO, London, 1984), 46, para. 8.7.
6. See R. E. Goodin, 'Exploiting a Situation and Exploiting a Person', in Andrew Reeve (ed.), *Modern Theories of Exploitation* (Sage Publications, London, 1987). This volume is an excellent source on the idea of exploitation. See also Joel Feinberg, 'Noncoercive Exploitation', in R. Sartorious (ed.), *Paternalism* (University of Minnesota Press, Minneapolis, 1983).
7. 'Exploiting a Situation', 179.
8. Goodin is to be commended here for it is always admirable in philosophy to attempt to find a unifying principle for disparate phenomena.
9. Anyone who has met Kim Cotton, who achieved a certain celebrity as the first commercial surrogate mother in the United Kingdom, will know that she was far from vulnerable and certainly not exploited.
10. Ibid. 189.
11. For a discussion of the impossibility of a clean hands policy see my *Violence & Responsibility* (Routledge & Kegan Paul, London, 1980).
12. I do not I hope beg any questions as to whether there are any valid reasons for such a judgement.
13. See Goodin, 'Exploiting a Situation', 188–9.

14. Feinberg, *Harmless Wrongdoing*, 206.
15. Here Feinberg seems to have brought his position more into line with that recommended in Ch. 4.
16. There is of course a real question as to how and to what extent an offer of money affects the autonomy of decision-making but I shall not address it here. I assume that financially attractive bargains may be struck autonomously and that some special reason why particular circumstances render the transaction non-autonomous must be shown.
17. See Hillel Steiner, 'Exploitation: A Liberal Theory Amended, Defended and Extended', in Reeve, *Modern Theories of Exploitation*, 132. Steiner's earlier essay is 'A Liberal Theory of Exploitation', in *Ethics*, 94 (1984), 225–41.
18. Though for a thorough analysis of just what this might be see Terrel Carver, 'Marx's Political Theory of Exploitation', in Reeve, *Modern Theories of Exploitation*.
19. See Steiner, 'A Liberal Theory of Exploitation', and David Miller, 'Exploitation in the Market', in Reeve, *Modern Theories of Exploitation*.
20. This seems to me parallel with the principle that supports parental custody of their genetic children. Someone should have custody of the children. There is no point in reallocating children to strangers unless their parents are manifestly unfit, so it is economical to leave children with their own parents. Add to this the fact that this is usually what the parents themselves want and a strong argument is established.
21. This is a very complicated issue. For one thing, to permit claims of adults on any and every neonate would make existence of the next generation precarious in the extreme. The teasing out of all the issues at stake here is most definitely a task for another occasion.
22. Of course there is a general point to be born in mind here from economic theory. It is that as you increase a person's wealth, the minimum price she will sell something for increases due to the diminishing marginal utility of cash for her; hence the value disparity required for exploitation diminishes in proportion to the alleviation of poverty. I am grateful to Hillel Steiner for pointing this out to me.
23. I give a somewhat fuller analysis of autonomy in my *The Value of Life* (Routledge & Kegan Paul, London, 1989), ch. 10 and of my views on children's rights in 'The Political Status of Children', in Keith Graham (ed.), *Contemporary Political Philosophy* (Cambridge University Press, Cambridge, 1982).
24. Here legislators would need to spell out clearly the meaning of 'available'. The idea obviously is that it is undesirable to do things to individuals who cannot consent and that all consenting sources should be utilized first.
25. See my *Violence & Responsibility*, ch. 5.

26. See the *Guardian* (5 Apr. 1990), 4.
27. See *Observer Magazine* (27 May 1990), 50.
28. I am assuming that it would be best to try to confine commercial transplants to citizens of the same community so that reciprocal benefit will be more closely perceived. The figure of 20,000 deaths is for the United States.

Chapter 7

1. I ignore the resource implications of all this.
2. Shapespeare, *The Merchant of Venice*, Act III, Scene i.
3. See Caroll Ezzell, 'First Ever Animal Patent Issues in United States', *Nature*, 332 (1987), 21.
4. I am told that to a biologist the terms chimera and hybrid are quite different organisms. The following distinctions would, I believe, be acceptable to a biologist. Chimeras have phenotypic diversity among their cells. They could be produced, for instance, by aggregating two embryos of different species, strain, or simply expressing different alleles of an enzyme, or by injecting a genomic DNA sequence plus carrier into a single cell of the blastocyst. In any one tissue (and not necessarily all of them) the resultant organism will show some cells of one origin/genotype and phenotype and some of another. A hybrid is an animal which is derived from the fertilization of the egg of one species with the sperm of a second and since the gametes were of two different species, the embryonic genotype *of every cell* is a mix of the two gametic (species) contributions leading to a resultant ('compromise') phenotype. A transgenetic creature will only *also* be a chimera if only some cells express the transgene.

   That said, biologists are not the arbiters of linguistic usage any more than compilers of dictionaries are. 'Chimera' is a word in common usage and I shall use it in a way that has been familiar since Bellerephon dispatched the paradigm.
5. These cases are usefully summarized in the *Guardian* (1 Aug. 1988).
6. Ibid.
7. See my *The Value of Life* (Routledge & Kegan Paul, 1985, 1989), ch. 1.
8. *Guardian*, 1 Aug. 1988.
9. Joseph Warkany, *Congenital Malformations* (Year Book Medical Publishers Inc., Chicago, 1971).
10. Ibid. 15.
11. E. E. Evans-Pritchard, *Nuer Religion* (OUP, Oxford, 1956).
12. See n. 20 below.
13. Warkany, *Congenital Malformations*, 15.

14. *First Born*, BBC TV, broadcast October and November 1988.
15. See Harris, *The Value of Life*, ch. 1.
16. I use the term 'pre-person' so as to make clear that I am not relying on any form of the potentiality argument.
17. In Ch. 2.
18. Mary Douglas, *Purity & Danger* (Routledge & Kegan Paul, London, 1970), 38.
19. Ibid. 40.
20. Ibid. 36.
21. Ibid. 39. See also E. E. Evans-Pritchard, *Nuer Religion.*
22. Jonathan Glover, *What Sort of People Should There Be?* (Penguin, Harmondsworth, 1984).
23. Ibid. 40.
24. Ibid. 41.
25. F. M. Cornford, *The Microcosmographia Academica* (Bowes & Bowes, London, 1908, repr. 1966).
26. Ibid. 23.
27. Ronald Dworkin, *Taking Rights Seriously* (Duckworth, London, 1977), 246, 247.
28. See Mary Warnock's remarks as reported in the *Sunday Correspondent* (22 Apr. 1990) (front page).
29. Perhaps only a British writer could think of including 'class' in this list.
30. Harris, *The Value of Life*, 8, pp. 166–73.
31. Edward Yoxen, *Unnatural Selection* (Heinemann, London, 1986).
32. Ibid. 114–15.
33. Ibid. 115.

Chapter 8

1. Karl Marx, *Theses on Feuerbach*, No. XI, in Lewis S. Fuer (ed.), *Marx and Engels* (Collins, Fontana, London, 1972).
2. Jonathan Glover *et al.*, *Fertility & The Family* (Fourth Estate, London, 1989).
3. David Suzuki and Peter Knutdson, *Genethics* (Unwin Hyman, London, 1989).
4. Ibid. 181.
5. Ibid. 202.
6. Ibid.
7. I have not been able to trace any others, but it would be rash indeed to state categorically that there are none.
8. Or conservatives, which amounts to the same thing.
9. I owe this point to Susan Kimber.

10. See CIBA Foundation Symposium 149, *Human Genetic Information: Science, Law and Ethics* (John Wiley, Chichester, 1990).
11. Ibid. 204.
12. F. M. Cornford, *The Microcosmographia Academica* (Bowes & Bowes, Cambridge, 1966), 26.
13. Suzuki and Knutdson, *Genethics*, 204, 205.
14. Ibid.
15. See my *The Value of Life* (Routledge & Kegan Paul, London, 1989), ch. 3.
16. Suzuki and Knutdson, *Genethics*, 205.
17. Shakespeare, *Hamlet*, Act III Scene i.
18. There is a parallel here between Suzuki and Knudtson's arguments and those of Lord Devlin arguing for the idea that the morality of a society is simply that of a particular group of people at a particular time. See Patrick Devlin, *The Enforcement of Morals* (Oxford University Press, Oxford, 1965) and the comprehensive demolition of these ideas by H. L. A. Hart in *Law Liberty & Morality* (Oxford University Press, Oxford, 1963) and by Ronald Dworkin in *Taking Rights Seriously* (Duckworth, London, 1977), ch. 10.
19. Suzuki and Knutdson, *Genethics*, 346.
20. Point mutations, alterations in single base pairs, if non-conservative, will give rise to a mutation in the whole gene. (Such point mutations will be conservative if the deletion of the base pairs is balanced a short distance down the chain by another deletion so that the triplets of base pairs still code the same amino acid sequence and no difference is made to the protein—hence 'conservative'). In the case of one amino acid change in the protein coded by that gene, if this change affects protein function in any way then this point mutation could have as much effect as much larger changes in the gene.
21. Suzuki and Knutdson, *Genethics*, 205.
22. Ibid. 205.
23. Ibid. 207.
24. See my *Violence & Responsibility* (Routledge & Kegan Paul, London, 1980).
25. See e.g. ibid. and Jonathan Glover, *Causing Death and Saving Lives* (Penguin, Harmondsworth, 1977).
26. BBC TV, *The Heart of the Matter*, on genetic engineering. Transmitted 22 Oct. 1989 (22.45 BBC 1). For quotations I am relying on the transcript of the programme prepared by the BBC. I have added what I hope is appropriate punctuation. Since the quotations are transcripts of 'live' unscripted interviews neither George Steiner nor Germaine Greer express themselves with their characteristic elegance.
27. See my *The Value of Life*, ch. 1.

28. *The Heart of the Matter.*
29. Ibid.
30. Ibid.
31. If indeed there is anything more nebulous than the concept of genius.
32. *The Heart of the Matter.* Germaine Greer does seem to have abandoned rationality on the subject of genetic engineering. In an ill-tempered piece in the *Independent Magazine* (8 July 1989) she first sloppily confuses thalassaemia with sickle cell disease when wrongly asserting the former's association with resistance to malaria. Then she sneeringly dismissed the idea that malaria might be cured by genetic engineering on the grounds that it is a parasitic infestation—the suggestion presumably being that you cannot combat parasites with genetic engineering. But she had already (though erroneously) noted that thalassaemia, an inherited condition, could confer immunity even to something caused by a parasitic infection.

Chapter 9

1. I must here again record a debt to my colleague Dr Susan Kimber, who has generously made many suggestions which have helped me to improve this chapter and hopefully also saved me from many errors. Complete salvation has doubtless eluded me!
2. See Mark W. J. Ferguson, 'Contemporary and Future Possibilities for Human Embryonic Manipulation', in Anthony Dyson and John Harris (eds.), *Experiments on Embryos* (Routledge & Kegan Paul, London, 1989), 14–15.
3. Ibid. This possibility we also examined in Ch. 5, but we will now be looking at a rather different dimension of the same problem.
4. There are of course many imponderables about this process and we should perhaps just mention two of them at this point. There is for example an important difference in what happens depending on at which stage of the gametes' development the insertion of the new gene is made. If the change is made early in gametogenesis (formation of the gametes) before the meiotic division (a mode of cell division which gives rise to the reproductive cells) then, if it has been inserted only into one of a pair of homologous chromosomes (identical chromosomes that pair during meiosis), only half the gametes will carry the gene.

   If on the other hand the insertion occurs into the haploid mature gametes then the transmission will depend on the success of the insertion technique. We will of course assume successful insertion and this could of course be checked, non-invasively, after completion of the attempt to see whether it has in fact been successful.

5. I ignore the boring 'philosophical' puzzle as to how many members there must be for a group of those members to constitute a 'breed'. Equally, I am not of course suggesting that the new breed would also be a new species in whatever technical meaning the word 'species' has.
6. Of course members of the new breed could pass on their genetic constitution with its advantages even if they did not procreate with other breed members. In subsequent generations 50% of children of any 'outbreeding' carrier would always inherit. Therefore, even if the individual did not breed within the carrier group transmission and spread of the gene and its effects would be fairly wide (assuming of course as we do that the gene is able to be expressed in the homozygous condition, i.e. is dominant). However, and I owe this suggestion, to Susan Kimber, it would not be satisfactory from the parents' point of view to know that only half their children (statistically) will be likely to benefit from the new technology. It would also open up interesting and perhaps serious inter-sibling problems.
7. Except that I am informed that current technology indicates that to protect against radiation damage or carcinogens it is likely that a number of genes coded for different ranges of repair enzymes would have to be inserted, making the process rather more cumbersome and perhaps problematic.
8. This vagueness need not worry us. All that is needed for present purposes is a powerful set of protections available to the new breed. Powerful enough to make them highly desirable and both personally and socially advantageous.
9. This perception may be apparent rather than real for very often industrial pollution is greater in less developed areas of the world. There are two sorts of reasons for this. One is that 'Western' multinational companies exploit third world countries and apply less stringent safety and pollution control standards than are insisted upon in the West. The second is that with so many other problems perception of the danger presented by pollution coupled with the difficulties of policing it may lead to more relaxed pollution control in the so-called third world.
10. They could of course simply modify their own children in the same ways as they themselves were modified but this would not 'maximize the advantage of their constituting a breed'.
11. See e.g. the discussion of these issues in Chapter 6.
12. I owe this suggestion to Peter Singer.
13. Again of course I am assuming that the new gene or genes are on both chromosomes and are dominant.
14. I am aware of course that work is also accelerating on creating new

breeds of plants which, like the new breed, are resistant to environmental pollutants. But here the same arguments apply. These will remain a small proportion of the world's living things.

15. Recently workers at the notorious British Nuclear Fuels plant at Sellafield have been advised not to have children because of the radiation damage suffered by workers at the plant. See Martin J. Gardner *et al.*, 'Results of Case-control Study of Leukaemia and Lymphoma among Young People near Sellafield Nuclear Plant in West Cumbria', in *British Medical Journal*, 300 (17 Feb. 1990). Reported in the *Guardian* (22 Feb. 1990). See also *Nature*, 343 (22 Feb. 1990). Although this is a case primarily affecting the work-force these sources show the danger is wider than this and affects the local community generally.
16. See e.g. Ronald Dworkin, *Taking Rights Seriously* (Duckworth, London, 1977); John Rawls, *A Theory of Justice* (Harvard University Press, Cambridge, Mass., 1971); Bruce Ackerman, *Social Justice in the Liberal State* (Yale University Press, New Haven, Conn., 1980).
17. See Ronald Dworkin's compelling argument for this in Dworkin, *Taking Rights Seriously*, ch. 6. Interestingly Dworkin's argument arises from his study of Rawls's theory of justice.
18. See ibid. 198, Robert Nozick, *Anarchy, State & Utopia* (Blackwell, Oxford, 1974), 33, and Jonathan Glover, *What Sort of People Should There Be?* (Penguin, Harmondsworth, 1984), 40–2.
19. See Ronald Dworkin's compelling elucidation of this distinction in *Taking Rights Seriously*, 227.
20. See e.g. my 'QALYfying the Value of Life', in *Journal of Medical Ethics* 13: 3 (Sept. 1987) and the contributions of Lockwood, Broome Harris, and Williams to S. Bell and M. Mendus (eds.), *Philosophy and Medical Welfare* (Cambridge University Press, Cambridge, 1988).
21. Thomas Hobbes, *The Leviathan* (1651). I am using Michael Oakshot's edition (Blackwell, Oxford, 1960), Part II, ch. 21, 'The obligation of subjects to the sovereign, is understood to last as long, and no longer, than the power lasteth, by which he is able to protect them.'
22. The nature of these opportunities has already been noted and will be of continuing interest as the discussion progresses.
23. See Chs. 3 and 4 above and of course Derek Parfit, 'Rights Interests and Possible People', in S. Gorovitz (ed.), *Moral Problems in Medicine* (Prentice-Hall, Engelwood Cliffs, NJ, 1976).
24. As we saw in Ch. 4 it might not be wrong to fail to protect an embryo from damage where a particular child would only exist if the protections were not applied.
25. George Orwell, *Nineteen Eighty-Four* (Penguin, Harmondsworth, 1954), 103, first published in 1949. One finds similar thoughts expressed by

'The Savage' in Aldous Huxley's *Brave New World*, published seventeen years before Orwell's book.

Chapter 10

1. Mark W. J. Ferguson, 'Contemporary and Future Possibilities for Human Embryonic Manipulation', in Anthony Dyson and John Harris (eds.), *Experiments on Embryos* (Routledge, London, 1990).
2. This particular issue I will not pursue further now. I hope to tackle it in another book, *Children's Liberation*, to be published by Routledge.
3. I am thinking here of course of the extensive recent interest in what is variously called 'preferential hiring', 'affirmative action', or 'compensatory discrimination'. In all cases a disadvantaged group, previously the subject of supposed unfair discrimination is deliberately preferred in employment, education, or some other important opportunity. There is an extensive literature of affirmative action which I will not detail here.
4. David Suzuki and Peter Knudtson, *Genethics* (Unwin Hyman, London, 1989), 161.
5. Ibid. 163.
6. Ferguson, 'Contemporary and Future Possibilities', 8.
7. Even supposing that one could, with a straight face, talk about a danger of such magnitude, however statistically remote, as 'effectively non-existent'.
8. See e.g. the excellent summary of much of the evidence in James Cutler and Rob Edwards, *Britain's Nuclear Nightmare* (Sphere Books, London, 1988) esp. ch. 6.
9. For those who favour conspiracy theories.
10. I do not of course beg any questions as to whether or not there are such things as rights. The language of rights is impossible to eradicate and here as always I just mean by it what it is on balance obligatory to do.
11. We have not examined dangers remote in time in any depth but they are as real. We should no more do damage that will accrue even in the remote future than we should do it to our neighbours.
12. The question of who precisely is to bear these costs, the employer or the community at large, is one we do not need to settle here.
13. Ronald Dworkin, *Taking Rights Seriously* (Duckworth, London, 1977), 199.
14. And I am of course by no means the first. See R. M. Hare, *Freedom and Reason* (Oxford University Press, Oxford, 1963) and John Rawls, *A Theory of Justice* (Harvard University Press, Cambridge, Mass., 1971), and, more recently, Jonathan Glover, *What Sort of People Should There Be?* (Penguin, Harmondsworth, 1984).

15. Of course this claim that it is worse to violate the equality principle is just that—a claim. The claim could also be defended by a sort of rule-utilitarian argument along the following lines: (1) the practice of non-discrimination, observance of the equality principle, produces on-balance net benefits, (2) particular cases of discrimination may appear to be net beneficial, but (3) particular breaches of the equality principle tend to undermine the practice as a whole, and (4) therefore, because of (1) particular breaches are ultimately not net beneficial and should be forbidden. There may of course be particular breaches that are *so beneficial* as to defeat this claim but these will be extreme cases.
16. See also my *The Value of Life* (Routledge & Kegan Paul, London, 1985) and my 'More & Better Justice' in S. Bell and M. Mendus (eds.), *Philosophy and Medical Welfare* (Cambridge University Press, Cambridge, 1988).
17. Related campaigns in the United Kingdom have involved the right of Sikhs to ignore legislation requiring motorcyclists to wear crash helmets and permit Sikhs to wear their traditional 'turban' only, despite obviously greater risks of injury.
18. Suzuki and Knudtson, *Genethics*, 175.

Chapter 11

1. Gosciny and Uderzo, *Astérix le gaulois* (Dargaud Éditeur, Neuilly-sur-Seine, 1970).
2. 'Virtually' now no longer meaning truly but 'almost truly' or 'not quite truly', the rather uglier locution 'veritably' must serve in its stead.
3. All life insurance and health insurance forms in the United Kingdom now contain questions not only as to HIV status but as to whether the candidate for insurance has been tested for HIV. As far as I know the deplorable consequences mentioned here have not as yet been documented or researched.
4. I have discussed this problem at length in my *The Value of Life* (Routledge & Kegan Paul, 1985, 1989), chs. 10 and 11.
5. It is a bit late to introduce the philosophical distinction between truthful and true statements, but for the record a 'truthful' statement is one made sincerely, by a person who believes it to be true. A 'true' statement on the other hand is of course simply one that is in fact true whether or not sincerely uttered. And while we are rehearsing distinctions 'libel' is of course written or otherwise published damaging falsehood and 'slander' is spoken damaging falsehood.
6. See Marjorie Sun, 'Scientists Settle Cell Line Dispute', *Science*, 220 (1983), 293–394.

7. Barbara J. Culliton, 'Mo Cell Case Has its First Court Hearing', in *Science*, 226 (1984), 813–14. See also Sandra Blakeslee, 'Patient Sues for Title to Own Cell Lines', *Nature*, 311 (1984), 198. I must thank Charles Erin for drawing these cases to my attention.
8. Blakeslee, 'Patient Sues for Title'.
9. For an excellent summary of the considerations surrounding the commercial exploitation of cell lines and the like see Norman H. Carey and P. E. Crawley, 'Commercial Exploitation of the Human Genome: What are the Problems?', in CIBA Foundation Symposium 149, *Human Genetic Information: Science, Law and Ethics* (John Wiley, Chichester, 1990), 133–48.
10. My colleague Margot Brazier was responsible for my knowing the, hopefully correct, legal formula for this wrong. If there is error it results from my improper understanding of her excellent tuition.
11. Culliton, 'Mo Cell Case', 814.
12. By 'circle of confidentiality' I mean the professional group who have responsibility for the immediate care of the patient and his records, whether that responsibility be contractual, voluntary, or statutory.
13. Hillel Steiner, *An Essay on Rights* (Basil Blackwell, Oxford, forthcoming), ch. 7. I am much indebted to Hillel for very generous comments on this chapter and for stimulating much of my interest in these questions. A further discussion of these issues will appear in his paper, 'The Fruits of Bodybuilders' Labour', in John Harris and Anthony Dyson (eds.), *Biotechnology & Ethics* (Routledge, London, forthcoming).
14. Again I must record Hillel Steiner's influence here and defer to his excellent discussion of these points in *An Essay on Rights*.
15. This point is Hillel Steiner's.
16. See e.g. my *The Value of Life*, ch. 11.
17. See *Prenatal Diagnosis and Genetic Screening*, A Report of the Royal College of Physicians (The Royal College of Physicians of London, Sept. 1989), 32.
18. Ibid. 28.
19. Ibid. 29.

Further Reading

The notes at the end of each chapter contain a comprehensive bibliography of the texts cited or referred to and many suggestions for further reading.

In the first section of further reading I list two sources for my own general philosophical position and a few good, and for the most part accessible, general introductions to ethics.

In the next section I list some (mostly readily available) books, both monographs and collections of essays, which deal with particular aspects of biotechnology, the law related to biomedicine and a few which deal with some of the ethical problems raised by biotechnology. This section also includes a few specially useful papers.

The final section is a select bibliography which cites other, mostly scientific, source material I have consulted but which is not necessarily included in the notes to each chapter.

1. Ethical foundations

Glover, Jonathan, *Causing Death and Saving Lives* (Pelican, Harmondsworth, 1977).

Harris, John, *Violence & Responsibility* (Routledge & Kegan Paul, London, 1980).

—— *The Value of Life* (Routledge & Kegan Paul, London, 1989).

Nagel, Thomas, *Mortal Questions* (Cambridge University Press, Cambridge, 1979).

Parfit, Derek, *Reasons & Persons* (Oxford University Press, Oxford, 1984).

Raz, Joseph, *The Morality of Freedom* (Oxford University Press, Oxford, 1986).

Singer, Peter, *Practical Ethics* (Cambridge University Press, Cambridge, 1979).

—— (ed.), *Applied Ethics* (Oxford University Press, Oxford, 1986).

2. Texts dealing with ethics, law and biotechnology

Books

Austin, C. R., and Short, R. V. (eds.), *Reproduction in Mammals* (Cambridge University Press, Cambridge, 1982).

Austyn, Jonathan M. (ed.), *New Prospects for Medicine* (Oxford University Press, Oxford, 1988).
Brazier, Margaret, *Medicine Patients and the Law* (Pelican, Harmondsworth, 1987).
Brown, T. A., *Gene Cloning: An Introduction* (Van Nostrand Reinhold, Wokingham, 1986).
CIBA Foundation Symposium 149, *Human Genetic Information: Science, Law & Ethics* (John Wiley, Chichester, 1990).
CIBA Foundation Symposium 130, *Molecular Approaches to Human Polygenic Disease* (John Wiley, Chichester, 1987).
CIBA Foundation Study Group, *Human Embryo Research: Yes or No?* (Tavistock, London, 1986).
Dawkins, Richard, *The Blind Watchmaker* (Penguin, Harmondsworth, 1988).
—— *The Selfish Gene* (Oxford University Press, Oxford, 1989).
Dunstan G. R., and Seller, Mary J. (eds.), *The Status of the Human Embryo* (King Edward's Hospital Fund for London, London, 1988).
Dyson, Anthony, and Harris, John (eds.), *Experiments on Embryos* (Routledge, London, 1990).
Edwards, Robert, *Life Before Birth* (Hutchinson, London, 1989).
—— and Purdy, J. (eds.), *Human Conception In Vitro* (Academic Press, London, 1981).
Fincham, J. R. S., and Ravetz, J. R., *Genetically Engineered Organisms* (Open University Press, Buckingham, 1991).
Ford, Norman M. *When Did I Begin?* (Cambridge University Press, Cambridge, 1988).
Glover, Jonathan, *What Sort of People Should There Be?* (Pelican, Harmondsworth, 1984).
—— *et al.*, *Fertility & The Family* (The Glover Report on Reproductive Technologies to the European Commission) (Fourth Estate, London, 1989).
Kennedy, Ian, *Treat Me Right* (Oxford University Press, Oxford, 1988).
Monk, Marilyn, and Surani, Azim, (eds.), *Genomic Imprinting* (The Company of Biologists Ltd., Cambridge, 1990).
Office of Technology Assessment, US Congress, *Impacts of Applied Genetics: Micro-organisms, Plants and Animals* (US Government Printing Office, Washington DC, 1981).
Oliver, Stephen G., and Ward, John M., *A Dictionary of Genetic Engineering* (Cambridge University Press, Cambridge, 1985).
President's Commission for the Study of Ethical Problems in Medicine and Biomedical and Behavioural Research, *Splicing Life: A Report on the Social and Ethical Issues of Genetic Engineering with Human Beings* (US Government Printing Office, Washington, DC, Nov. 1982).

The Royal College of Physicians of London, Report: *Prenatal Diagnosis and Genetic Screening* (London, 1989).
Slack, J. M. W., *From Egg to Embryo* (Cambridge University Press, Cambridge, 1983).
Suzuki, David, and Knudtson, Peter, *Genethics* (Unwin Hyman, London, 1988).
Warnock, Mary, *A Question of Life* (Basil Blackwell, Oxford, 1984).
Weatherall, D. J., *The New Genetics & Clinical Practice*, 2nd edition (Oxford University Press, Oxford, 1985).
Yoxen, Edward, *The Gene Business* (Pan Books, London, 1983).
—— *Unnatural Selection* (Heinemann, London, 1986).

Papers

Handiside, A. H., Kontogianni, E. H., Hardy, K., and Winston, R. M., 'Pregnancies from Biopsied Human Preimplantation Embryos Sexed by Y-specific DNA Amplification', *Nature*, 344 (19 Apr. 1990), 768–70.
Harris, John, 'In Vitro Fertilization: The Ethical Issues', in *Philosophical Quarterly*, 33: 132 (1983), 225.
Jeffrey's, A. J., Wilson, V., Thein, S. L., 'Hypervariable "Minisatellite" Regions in Human DNA', *Nature*, 314 (1985), 67–73.
Johnson, Martin, 'The Onset of Human Identity and its Relationship to Legislation Concerning Research on Human Embryos', *British Journal of Family Planning* (1988).
Rowland, Robyn, 'Making Women Visible in the Embryo Experimentation Debate', *Bioethics*, 1: 2 (1987).
Warnock, Mary, 'In Vitro Fertilization: The Ethical Issues II', in *Philosophical Quarterly*, 33: 132 (1983), 239.
—— 'Do Human Cells Have Rights?' *Bioethics*, 1: 1 (Jan. 1987).
—— 'The Good of the Child', *Bioethics*, 1: 2 (Apr. 1987).

3. Select Bibliography

Amernick, B. 'Essentials of Patent Law', *J. Natl. Cancer Inst.*, 81: 19, (4 Oct. 1989), 1450–4.
Anderson, D., and Cuthbertson, W. F., 'Safety Testing of Novel Food Products Generated by Biotechnology and Genetic Manipulation', *Biotechnol. Genet. Eng. Rev.*, 5 (1987), 369–95.
Anderson, W. F. 'Human Gene Therapy: Why Draw a Line?', *J. Med. Philos.*, 14: 6 (Dec. 1989), 681–93.
Annas, G. J. 'Who's Afraid of the Human Genome', *Hastings Cent. Rep.*, 19: 4 (July–Aug. 1989), 19–21.
Arnheim, N. and Erlich, H. A. 'Commercial Uses of Recombinant DNA

Technology in Human Genetic Disease', *Prog. Med. Genet.*, 7 (1988), 195–219.

Autrup, H. 'Carcinogen Metabolism in Cultured Human Tissues and Cells', *Carcinogenesis*, 11: 5 (May 1990), 707–12.

Barnhart, B. J. 'The Human Genome Project: A DOE Perspective', *Basic Life Sci.*, 46 (1988), 161–6.

Bogard, W. C. Jr., Dean, R. T., Deo, Y., Fuchs, R., Mattis, J. A., McLean, A. A. and Berger, H. J. 'Practical Considerations in the Production, Purification, and Formulation of Monoclonal Antibodies for Immunoscintigraphy and Immunotherapy', *Semin. Nucl. Med.* 19: 3 (July 1989), 202–20.

Borden, E. C. and Sondel, P. M. 'Lymphokines and Cytokines as Cancer Treatment: Immunotherapy Realized', *Cancer*, 65: 3 (suppl.) (1 Feb. 1990), 800–14.

Botkin, J. R. 'Ethical Issues in Human Genetic Technology, *Pediatrician (Switzerland)* 17: 2 (1990), 100–7.

Brahams, D. 'Human Genetic Information: The Legal Implications', *Ciba Found Symp.*, 149 (1990), 111–19.

Bratanov, K. and Vulchanov, V. H. 'Reproductive Immunology: A Prospective Marginal Field of Biomedical Sciences', *Am. J. Reprod. Immunol. Microbiol.*, 10: 3 (March 1986), 68–73.

Carey, N. H. 'Commercial Exploitation of the Human Genome: What are the Problems?', *Ciba Found Symp.*, 149 (1990), 133–43.

Carey, N. J. 'Production and Use of Therapeutic Agents', *Br. Med. J. [Clin. Res.]*, 295: 6603 (10 Oct. 1987), 907–8.

Carney, W. P., 'Application of Biotechnology Methods to the Study of Trematodes', *Southeast Asian J. Trop. Med. Public Health*, 19: 1 (Mar. 1988), 59–69.

Chang, T. M., 'Artificial Cells in Medicine and Biotechnology', *Appl. Biochem. Biotechnol.*, 10 (1984), 5–24.

—— 'Applications of Artificial Cells in Medicine and Biotechnology', *Biomater. Artif. Cells Artif. Organs*, 15: 1 (1987), 1–20.

—— 'The Role of Biotechnology in Bioartificial or Hybrid Artificial Cells and Organs [editorial]', *Biomater. Artif. Cells Artif. Organs*, 16: 4 (1988), pp. v–vi.

Charlesworth, M. 'Human Genome Analysis and the Concept of Human Nature', *Ciba Found Symp.*, 149 (1990), 180–9.

Clayton, R. N., 'The Molecular Biology of the Ovary and Testis', *Baillieres Clin. Endocrinol. Metab.*, 2: 4 (Nov. 1988), 987–1002.

Conway de Macario, E., and Macario, A. J., 'Monoclonal Antibodies for Bacterial Identification and Taxonomy: 1985 and Beyond', *Clin. Lab. Med.*, 5: 3 (Sept. 1985), 531–44.

Corcoran, E., 'A Tiny Mouse Came Forth', *Sci. Am.*, 260: 2 (Feb. 1989), 73.

—— 'Patent Medicine', *Sci. Am.*, 259: 3 (Sept. 1988), 128–30.
—— 'Gene Therapy in Gestation', *Sci. Am.*, 259: 5 (Nov. 1988), 135–6.
Cross, J. H., 'Biotechnology Research on Cestodes in the Philippines, Malaysia, and Indonesia', *Southeast Asian J. Trop. Med. Public Health*, 19: 1 (Mar. 1988), 41–5.
Crouch, M. L. 'Regulating Recombinant DNA Biologics', *Arzneimittelforschung*, 38: 7 (July 1988), 947–9.
De Jong, O. J., 'Ethical Aspects of Biotechnology', *Tijdschr Diergeneeskd*, 112: 2 (15 Jan. 1987), 84–8.
Delisi, D., 'Overview of Human Genome Research', *Basic Life Sci.*, 46 (1988), 5–10.
Dickman, S., 'Genetic Engineering: New Law is Overdue [news]', *Nature*, 342: 6247 (16 Nov. 1989), 218.
—— 'Genetic Engineering: New Laws Need Changes Made [news]', *Nature*, 343: 6256 (25 Jan. 1990), 298.
Dickson, D., 'Genome Project Gets Rough Ride in Europe [news]', *Science*, 243: 4891 (3 Feb. 1989), 599.
—— 'Europe Tries to Untangle Laws on Patenting Life [news]', *Science*, 243: 4894 pt. 1 (24 Feb. 1989), 1002.
Duffus, J. H., and Brown, C. M., 'Health Aspects of Biotechnology', *Ann. Occup. Hyg.*, 29: 1 (1985), 1–12.
Dzau, V. J., Paul, M., Nakamura, N., Pratt, R. E., and Ingelfinger, J. R., 'Role of Molecular Biology in Hypertension Research: State of the Art Lecture', *Hypertension*, 13: 6 pt. 2 (June 1989), 731–40.
Fields, B. N., and Chanock, R. M., 'What Biotechnology Has to Offer Vaccine Development', *Rev. Infect. Dis.*, 11: Suppl. 3 (May–June 1989), S519–23.
Fox, J. L., and Klass, M. 'Antigens Produced by Recombinant DNA Technology', *Clin. Chem.*, 35: 9 (Sept. 1989), 1838–42.
Friedmann, T. 'Applications and Implications of Genome Related Biotechnology', *Basic Life Sci.*, 46 (1988), 149–59.
Gelberman, R. H., and Dimick, M. P., 'The Biotechnology of Hand and Wrist Implant Surgery and Rehabilitation', *J. Rheumatol.*, 14: Suppl. 15 (Aug. 1987), 53–61.
Gershon, D. 'Genentech Shares: Allegation of Insider Dealing [news]', *Nature*, 345: 6271 (10 May 1990), 102.
Gilbert, W. 'Human Genome Sequencing', *Basic Life Sci.*, 46 (1988), 29–36.
Greenstein, R. L., 'Controlling the Bugs: The First Decade in the Regulation of Biotechnology', *Appl. Biochem. Biotechnol.*, 11: 6 (Dec. 1985), 489–506.
Grisolia, S., 'Mapping the Human Genome', *Hastings Cent. Rep.*, special suppl. (July–Aug. 1989), 18–19.
Guillou, P. J., 'Potential Impact of Immunobiotechnology on Cancer Ther-

apy', *Br. J. Surg.*, 74: 8 (Aug. 1987), 705–10.

Hadsell, R. M., 'Discover it Today, Tax it Tomorrow: Basic Biomedical Research Spawns Biotech Industry [news]', *J. Natl. Cancer Inst.*, 81: 5 (1 Mar. 1989), 322–6.

Hannah, H. W., 'Biotechnology and the Veterinarian: Some Legal Considerations', *J. Am. Vet. Med. Assoc.*, 194: 7 (1 Apr. 1989), 890–1.

Hashimoto, S. 'Recent Advances in Tumor-seeking Agents: Radioimmunoimaging of Cancers Using Monoclonal Antibodies', *Gan. To Kagaku Ryoho*, 15: 4, pt. 2–1 (Apr. 1988), 866–72.

Hill, C. R., 'Large-scale Manufacture of Monoclonal Antibodies for Use in Humans', *Biochem. Soc. Trans.*, 19: 2 (Apr. 1990), 245–7.

Hilleman, M. R., 'Newer Directions in Vaccine Development and Utilization', *J. Infect. Dis.*, 151: 3 (Mar. 1985), 407–19.

Hodgins, D. S., 'Life Forms Protectable as Subjects of US Patents: Microbes to Animals (Perhaps)', *Appl. Biochem. Biotechnol.*, 16 (Sept.–Dec. 1987), 79–93.

Hodgkin, P. and Yoxen, E., 'Biotechnology and General Practice. 2: Beyond the Technology—Social and Ethical Problems', *J. R. Coll. Gen. Pract.*, 35: 280 (Nov. 1985), 527–31.

Hoffman, T., Fratantoni, J., and Murano, G., 'Clinical Use of Biologicals Produced in Continuous Cell Lines', *Dev. Biol. Stand.*, 70 (1989), 211–14.

Hood, L. 'Biotechnology and Medicine of the Future', *J. of Am. Med. Assoc.*, 259: 12 (25 Mar. 1988), 1837–44.

Huber, B. E., 'Therapeutic Opportunities Involving Cellular Oncogenes: Novel Approaches Fostered by Biotechnology', *FASEB J.*, 3: 1 (Jan. 1989), 5–13.

International Congress on New Trends in Nephrology, Dialysis and Transplantation, Bologna, 14–16 April, 'Biotechnology in Renal Replacement Therapy', *Contrib. Nephrol.*, 70 (1989), 1–346.

Iritani, A., 'Current Status of Biotechnological Studies in Mammalian Reproduction', *Fertil. Steril.*, 50: 4 (Oct. 1988), 543–51.

Jacobsen, H. and Kirchner, H. 'Interferons: Biologic Principles and Clinical Uses', *Klin Padiatr.*, 197: 4 (July–Aug. 1985), 263–7.

James, K. 'Human Monoclonal Antibody Technology: Are its Achievements, Challenges, and Potential Appreciated?', *Scand. J. Immunol.*, 29: 3 (Mar. 1989), 257–64.

Janssen, D. B., and Witholt, B. 'Developments in Biotechnology of Relevance to Drinking Water Preparation', *Sci. Total Environ.*, 47 (Dec. 1985), 121–35.

Jones, E. H., 'Recombinant Human Erythropoietin', *Am. J. Hosp. Pharm.*, 46: 11, suppl. 2 (Nov. 1989), S20–3.

Kahan, J. S., and Gibbs, J. N., 'Food and Drug Administration Regulation of Medical Device Biotechnology, and Food and Food Additive Biotech-

nology', *Appl. Biochem. Biotechnol.*, 11: 6 (Dec. 1985), 507–16.
Kaufmann, S. H., 'Leprosy and Tuberculosis Vaccine Design', *Trop. Med. Parasitol.*, 40: 3 (Sept. 1989), 251–7.
Kingsbury, D. T., 'Balancing Regulatory Control, Scientific Knowledge, and Public Understanding', *Basic Life Sci.*, 45 (1988), 341–50.
Knezevic, S., 'Science and Bioethics: Guidelines, Dilemmas and the Constant Postponement of Decisions', *Lijec Vjesn*, 111: 12 (Dec. 1989), 427–31.
Kolb, E., 'New Knowledge and Aspects in the Use of Biotechnology', *Z. Gesamte Inn. Med.*, 41: 9 (1 May 1986), 249–55.
Kupper, H., 'Biotechnological Approach to a New Foot-and-mouth Disease Virus Vaccine', *Biotechnol. Genet. Eng. Rev.*, 1 (1984), 223–59.
Liew, F. Y., 'Biotechnological Trends towards Synthetic Vaccines', *Immunol. Lett.*, 19: 3 (Nov. 1988), 241–4.
Loeffler, K., 'Ethical Demands in Animal Production', *Berl. Munch. Tierarztl. Wochenschr.*, 102: 12 (1 Dec. 1989), 397–400.
McConnell, J. B., 'Whence We've Come, Where We're Going, How We're Going to Get There', *Basic Life Sci.*, 46 (1988), 1–4.
McLeod, D. C., 'Biotechnology: Product Development and Evolving Patent Law (editorial)', *DICP*, 23: 7–8 (July–Aug. 1989), 605–6.
Malik, V. S., 'Biotechnology: The Golden Age', *Adv. Appl. Microbiol.*, 34 (1989), 263–306.
Marwick, C., 'Technology, Cost, Cooperation, Ethics Challenges Face Genome Mapping Plan [news]', *J. of Am. Med. Assoc.*, 262: 23 (15 Dec. 1989), 3247.
Medvedev, Z. A., 'The Past and the Future of Experimental Gerontology', *Arch. Gerontol. Geriatr.*, 9: 3 (Nov.–Dec. 1989), 201–13.
Mooney, M. A., and Norby, S. 'Cost Benefit Analysis and Ethics of Prenatal Diagnosis, *Ugeskr Laeger*, 151: 44 (30 Oct. 1989), 2868–71.
Murray, K., Stahl, S., and Ashton-Rickardt, P. G., 'Genetic Engineering Applied to the Development of Vaccines', *Philos. Trans. R. Soc. Lond. [Biol.]*, 324: 1224 (31 Aug. 1989), 461–76.
Nature, 'Another Report Smiles on Human Genome Sequencing Project [news]', *Nature*, 332: 6167 (28 Apr. 1988), 769.
Nethersell, A. B., 'Biological Modifiers and their Role in Cancer Therapy', *Ann. Acad. Med. Singapore*, 19: 2 (Mar. 1990), 223–34.
Newmark, P., 'Europe Evaluates its Four-year Plans in Biotechnology [news]', *Nature*, 335: 6191 (13 Oct. 1988), 579.
Nolan, K., and Swensen, S., 'New Tools, New Dilemmas: Genetic Frontiers', *Hastings Cent. Rep.*, 18: 5 (Oct.–Nov. 1988), 40–6.
Nosik, D. N., Korsum, N. S., and Novokhatskii, A. S., 'Testing of Monoclonal Antibodies to Human Interferon', *Vopr. Virusol.*, 30: 5 (Sept.–Oct. 1985), 600–2.
Office of Health and Environmental Research, US Department of Energy,

'Roundtable Forum: The Human Genome Initiative: Issues and Impacts', *Basic Life Sci.*, 46 (1988), 93–109.

O'Hagan, D. T., Palin, K. J., and Davis, S. S., 'Intestinal Absorption of Proteins and Macromolecules and the Immunological Response', *Crit. Rev. Ther. Drug Carrier Syst.*, 4: 3 (1988), 197–220.

Overby, L. R., 'Challenges and Opportunities in Biotechnology', *Prog. Clin. Biol. Res.*, 182 (1985), 425–43.

Ozkan, A. N., Hoyt, D. B., and Ninnemann, J. L., 'Generation and Activity of Suppressor Peptides Following Traumatic Injury', *J. Burn Care Rehabil.*, 8: 6 (Nov.–Dec. 1987), 527–30.

Parkman, P. D., and Hopps, H. E., 'Viral Vaccines and Antivirals: Current Use and Future Prospects', *Ann. Rev. Public Health*, 9 (1988), 203–21.

Pinkert, C. A., Dyer, T. J., Kooyman, D. L. and Kiehm, D. J., 'Characterization of Transgenic Livestock Production', *Domest. Anim. Endocrinol.*, 7: 1 (Jan. 1990), 1–18.

Roberts, L., 'Genome Project: An Experiment in Sharing [news]', *Science*, 248: 4958 (25 May 1990), 953.

Rosenberg, S. A., 'The Development of New Immunotherapies for the Treatment of Cancer Using Interleukin-2: A Review', *Ann. Surg.*, 208: 2 (Aug. 1988), 121–35.

Sauer, F. and Hankin, R., 'Rules Governing Pharmaceuticals in the European Community', *J. Clin. Pharmacol.*, 27: 9 (Sept. 1987), 639–46.

Schoepke, H. G., 'The Prospective of the Biotechnology Industry on Changing Technology in Pathology', *Arch. Pathol. Lab. Med.*, 111: 7 (July 1987), 596–600.

Sorensen, S. A., 'Ethical Aspects in the Diagnosis of Hereditary Diseases, with Special Reference to Huntington's Chorea', *Nord Med.*, 105: 1 (1990), 2–4.

Stewart, H. J., Jones, D. S., Pascall, J. D., Popkin, R. M., and Flint, A. P. 'Drug Delivery Systems in Biotechnology', *Artif. Organs*, 12: 3 (June 1988), 248–51.

Stadler, R., Mayer-da-Silva, A., Bratzke, B., Garbe, C., and Orfanos, C., 'Interferons in Dermatology', *J. Am. Acad. Dermatol.*, 20: 4 (Apr. 1989), 650–6.

Sugawara, S., 'Micromanipulation of Mammalian Embryos', *J. Toxicol. Sci.*, 13: 4 (Nov. 1988), 279–86.

Sullivan, J. B. Jr., 'Immunotherapy in the Poisoned Patient: Overview of Present Applications and Future Trends', *Med. Toxicol.*, 1: 1 (Jan–Feb. 1986), 47–60.

Tami, J. A., Parr, M. D., and Thompson, J. S., 'The Immune System', *Am. J. Hosp. Pharm.*, 43: 10 (Oct. 1986), 2483–93.

Urushizaki, I. 'Recent Trends in Cancer Treatment Using Cytokines', *Gan To Kagaku Ryoho*, 16: 4, pt. 2–1 (Apr. 1989), 897–905.

Watson, J. D. and Jordan, E., 'The Human Genome Program at the National Institute of Health', *Geonomics*, 5: 3 (Oct. 1989), 654–6.

Werner, R. G., 'Biotechnical Production of Drugs with Special Reference to Tissue-type Plasminogen Activator', *Klin. Wochenschr.*, 66 suppl. 12 (1988), 24–32.

Westphal, H., 'Transgenic Mammals and Biotechnology', *FASEB. J.*, 3: 2 (Feb. 1989), 117–20.

Wirth, D. F., Rogers, W. O., Barker, R. Jr., Dourado, H., Suesebang, L., and L'Albuquerque, B., 'Leishmaniasis and Malaria: New Tools for Epidemiologic Analysis', *Science*, 234: 4779 (21 Nov. 1986), 975–9.

Young, F. E., 'DNA Probes. Fruits of the New Biotechnology', *J. of Am. Med. Assoc.*, 258: 17 (6 Nov. 1987), 2404–6.

Zilz, D. A., 'Harvey A. K. Whitney lecture. Interdependence in Pharmacy: Risks, Rewards, and Responsibilities', *Am. J. Hosp. Pharm.*, 47: 8 (Aug. 1990), 1759–65.

Zimmerli, W. C., 'Who Has the Right to Know the Genetic Constitution of a Particular Person?', *Ciba Found Symp.*, 149 (1990), 93–102.

Index